Abacus

For at least 2,300 years people have used the device we know as "Abacus" to compute mathematical problems.

 Some researchers say "the abacus is probably derived from an ancient counting board that was spread with sand and marks in the sand were used to count and calculate. The sand may have been synonymous with dust, in some dialects, and the Greek word Abax evolved into the Latin word Abacus.

The Abacus is not a necessary accounting tool in today's world, but the use of the Abacus is part of many cultures to aid in the development of non written, and non-verbal, calculating.

There are several versions of Abaci used by people all over the world. The basic design is a rectangular frame

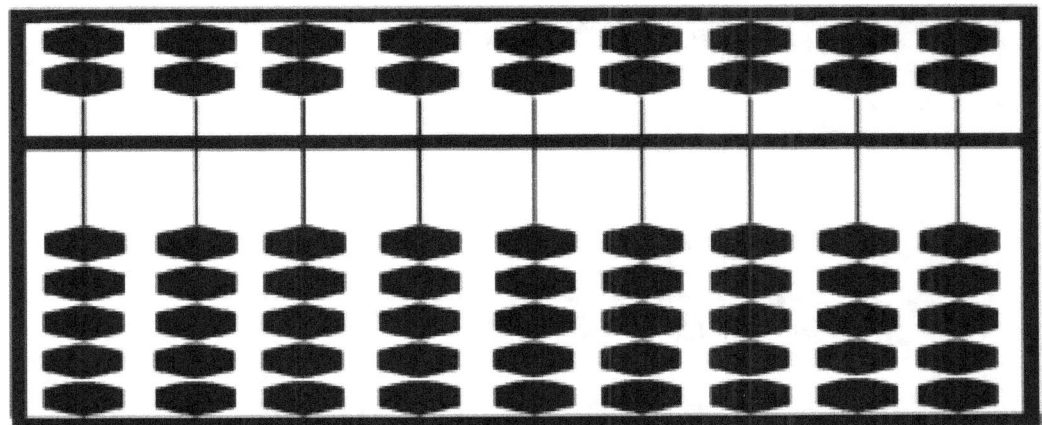

with several vertical rods or wires, upon which are mounted beads.

The beads are divided into upper and lower parts of the rods by a bar running horizontally across the frame. Generally, the beads above the bar are called "Heaven Beads" and those below the bar are called "Earth Beads".

There are several designs for an abacus in different cultures around the world. One version of a Russian abacus has rods running horizontally and beads of two or more colors, so no bar is necessary.

The Japanese abacus may have one bead above and four beads below the bar, on each rod.

In this tutorial I will use the Chinese Abacus with thirteen, or less, vertical rods or wires, each of which holds two beads in the top deck and five beads in the lower deck.

A number is expressed by pushing a bead, or beads, against the bar.

If possible you should use a Chinese Abacus, (Suanpan), with this book. There is no substitute for hands-on practice. Some public Libraries may have an abacus for check out. An abacus may be purchased online for less than fifteen dollars. Some abaci are antiques and many people have extensive collections of them.

These instruments are used even today in remote areas of Asia, Europe, India, and Middle East where electric systems are difficult to maintain or non-existent. It is not a bad idea to learn how to operate more than one style of abacus.

Let the learning begin.

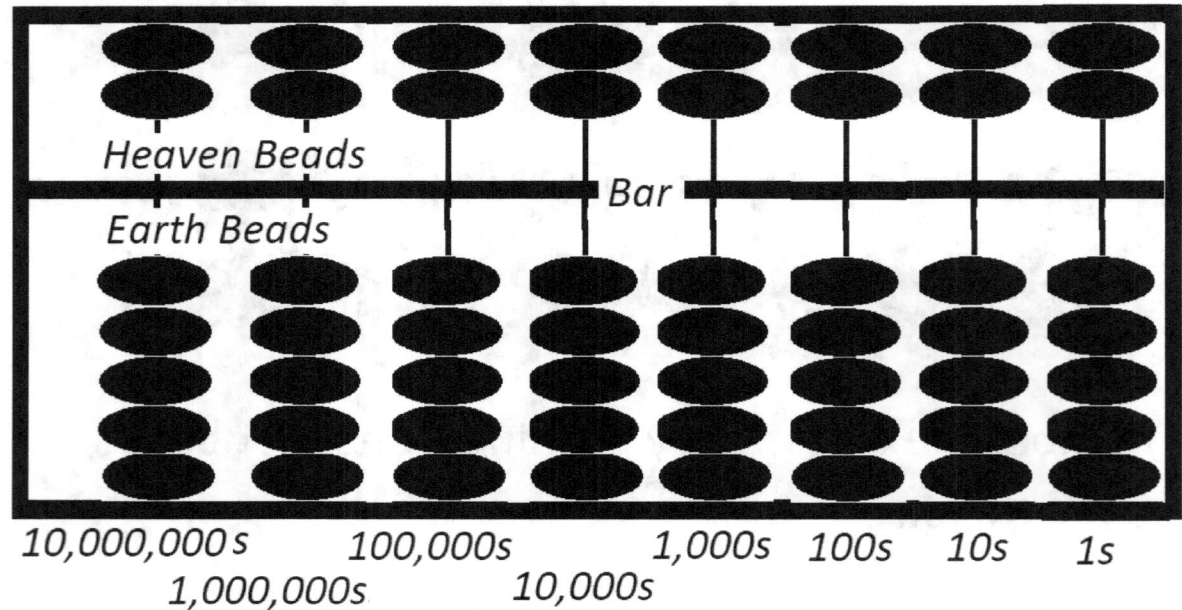

These are the place values for each column

Here is a basic Chinese Abacus.

Each column is holding beads, in the bottom and top decks, worth 10 times the beads on the adjacent column, to the right. One Heaven Bead is worth five Earth beads, on the same column. The following example shows the numbers 6 on column B, 1 on column C, and to the right, 5 on column G. Here we are disregarding place value.

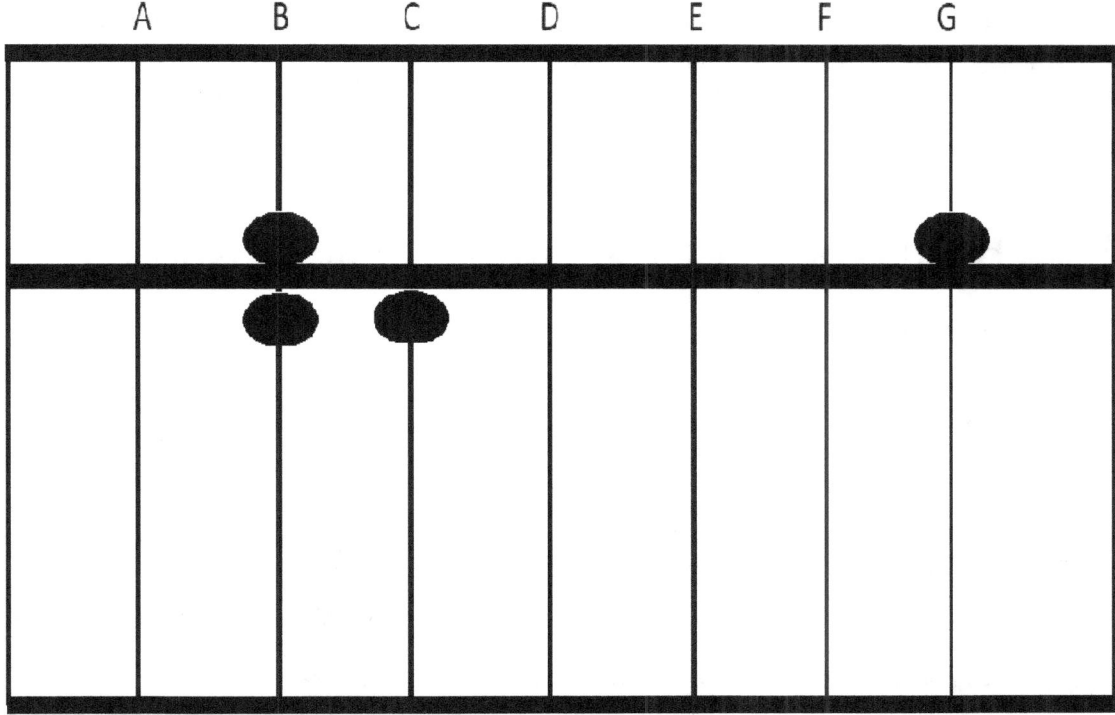

An abacus may have any number of columns, colors, or bead values, as long as there is a consistent relationship between beads, columns and place values.

As we go right to left, the place values of the columns we will use are, 1s, 10s, 100s, 1000s, and so on.

If we read the above as all one number with the most right column representing units, then we have the number 6 10,005.

To place all the beads in a starting position, we lay the frame flat and then tilt it toward us. This brings all the Earth Beads to the bottom, and all the Heaven Beads to the bar. We run a finger along the top of the bar to push all Heaven Beads to the top of the frame, away from the bar.

With no beads touching the bar, the value displayed is zero, or starting position.

Look at the rightmost column. Move the Earth Beads

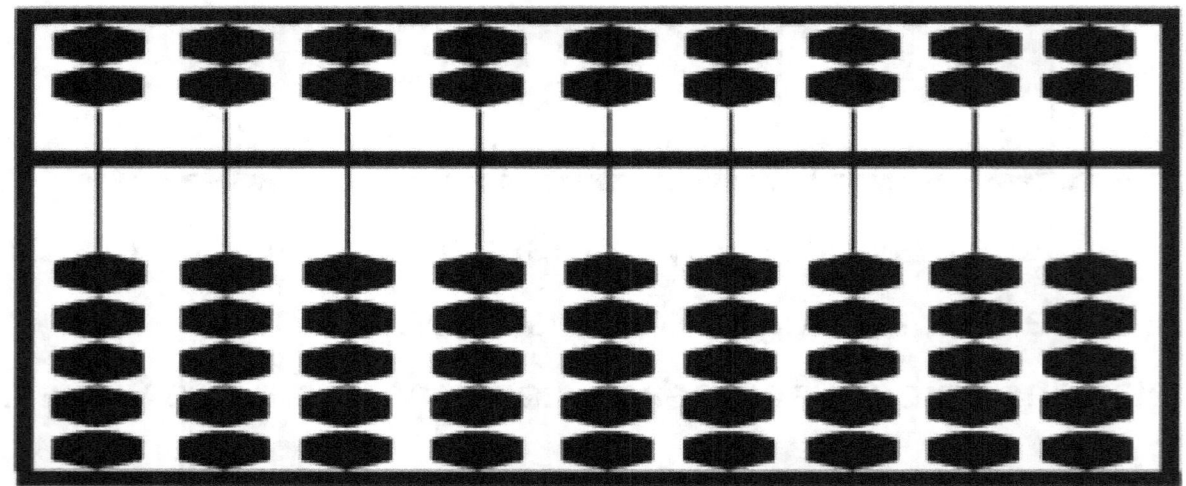

up toward the bar as you count them 1, 2, 3, 4, 5.

The number at the bar is now 5.

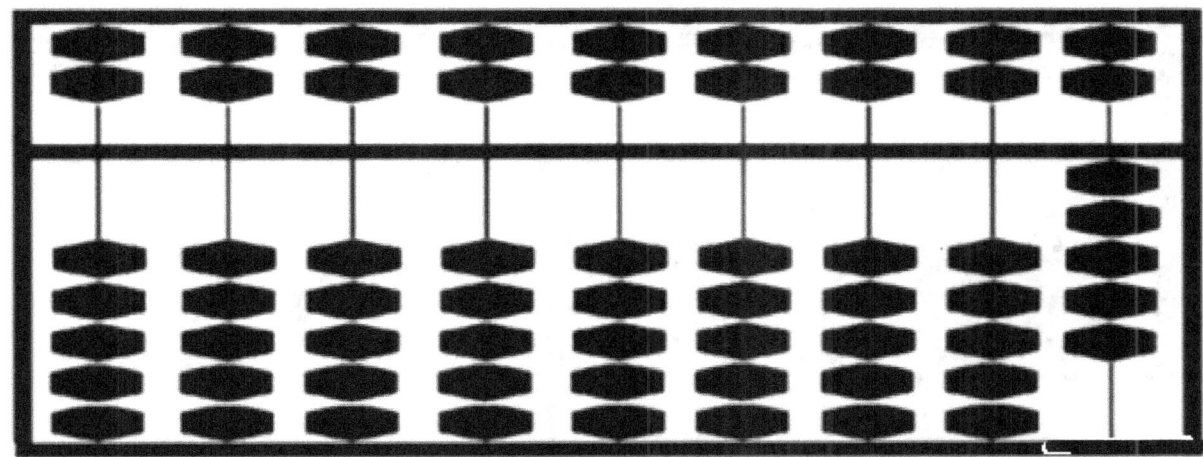

Now, we want to keep counting, so we move 1 heaven bead, on the same column, to the bar; this replaces the 5 Earth Beads which are moved back to the bottom.

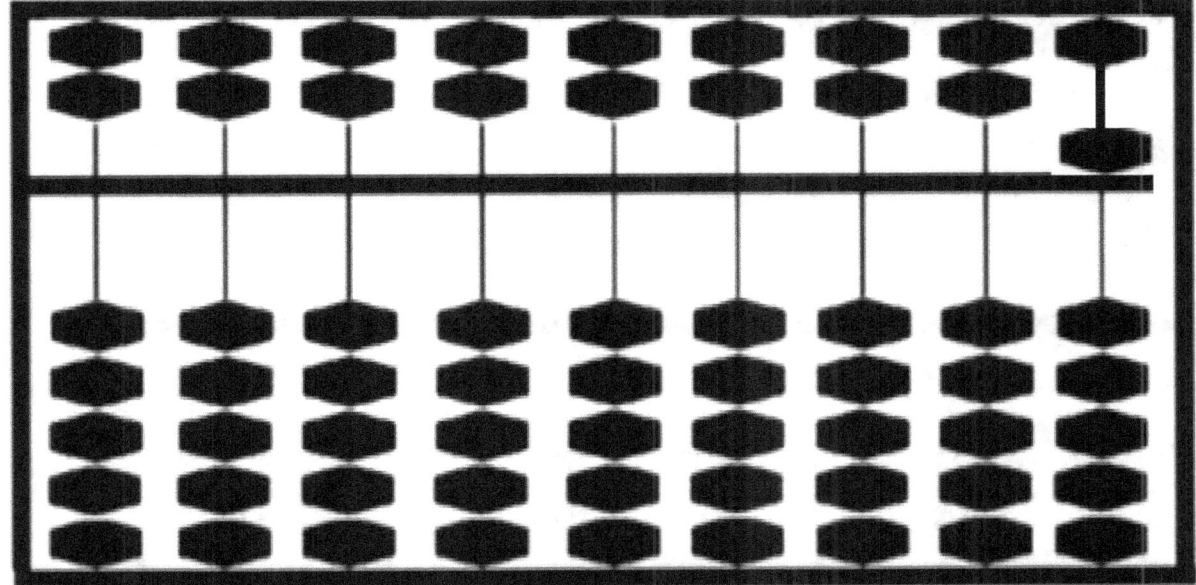

Let us continue to count the same beads, 6, 7, 8, 9, 10.

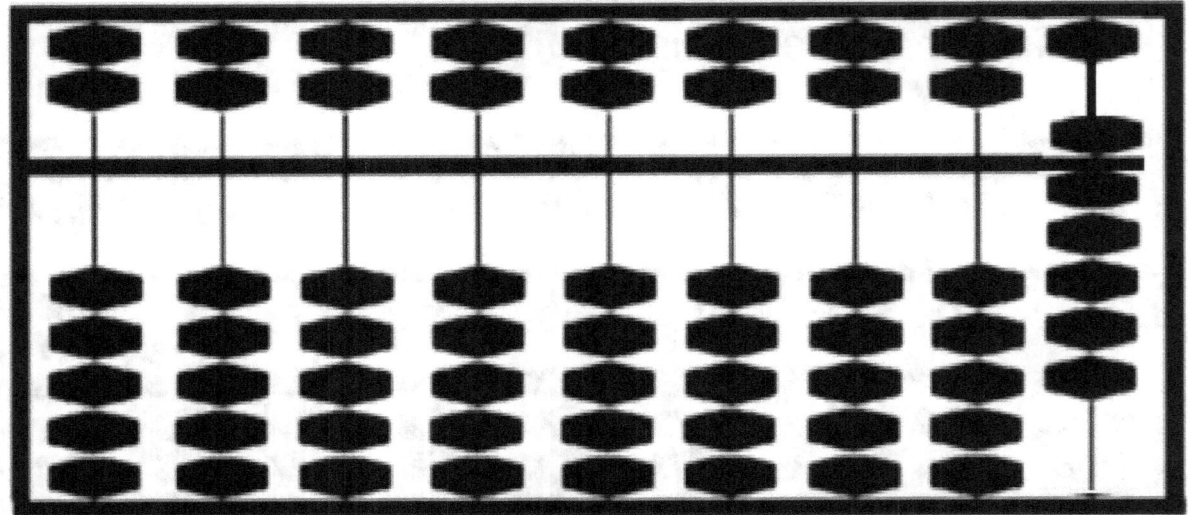

Again, we are out of units to count, so we move a second Heaven Bead toward the bar on the 1s column, to replace the 5 Earth Beads.

Now we can count the Earth Beads a third time, 11, 12, 13, 14, 15.

There are 2 Heaven beads (5 each), and 5 Earth Beads (1 each) at the bar for a total number of 15 at the bar.

In order to keep counting we go to the next column, to the left. We move one Earth Bead, on that column, up to the bar. 1 Earth Bead is equal to two Heaven Beads on the next column to the right.

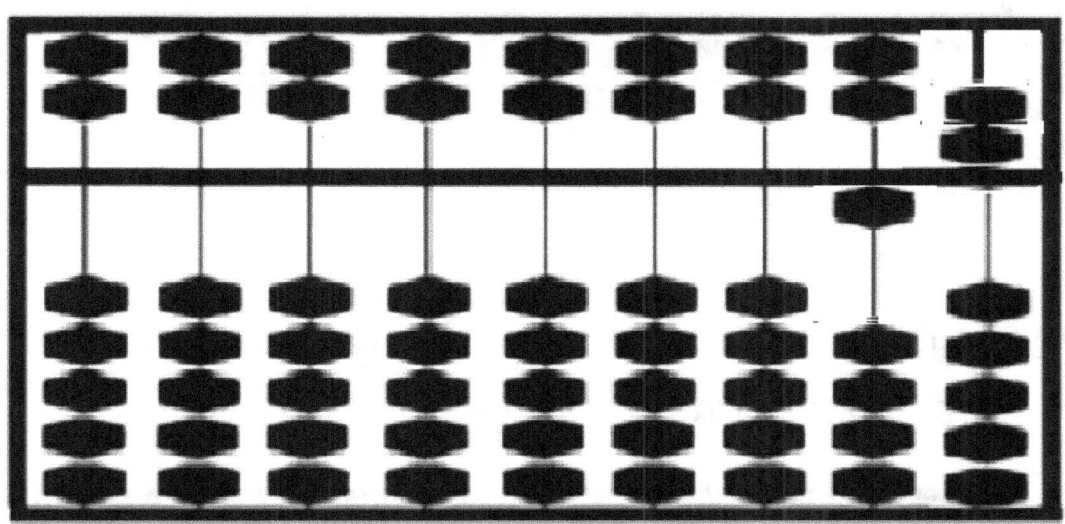

Every time we create a column with two Heaven Beads at the bar, we go to the next column on the left, and replace them with an Earth Bead which is the same value as the two Heaven Beads on the right.

This replaces the two Heaven Beads at the bar! The replaced beads are moved back to starting position away from the bar.

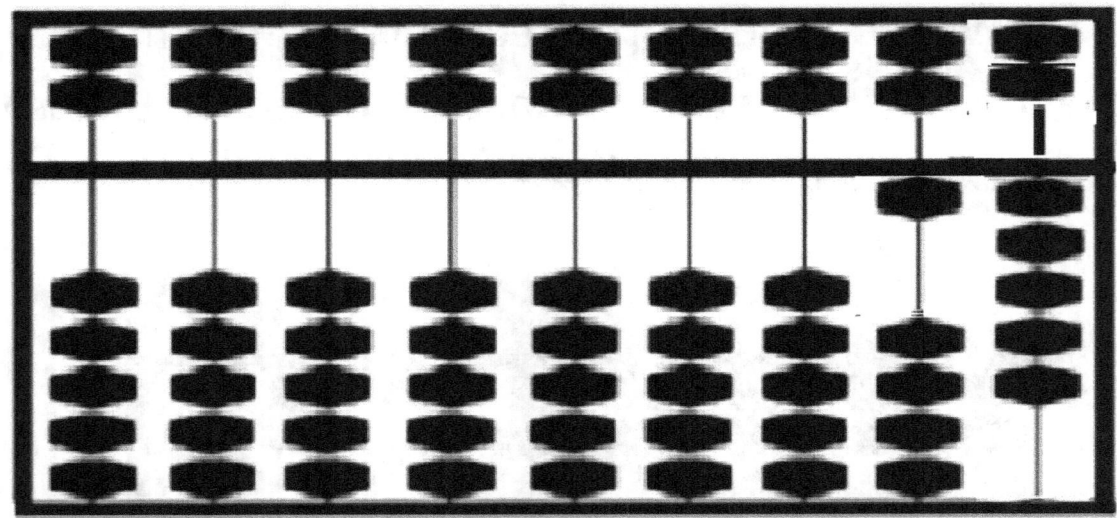

Now we can replace the Earth Beads, on the first column, again with a Heaven Bead and move the five Earth Beads away from the bar on the first column.

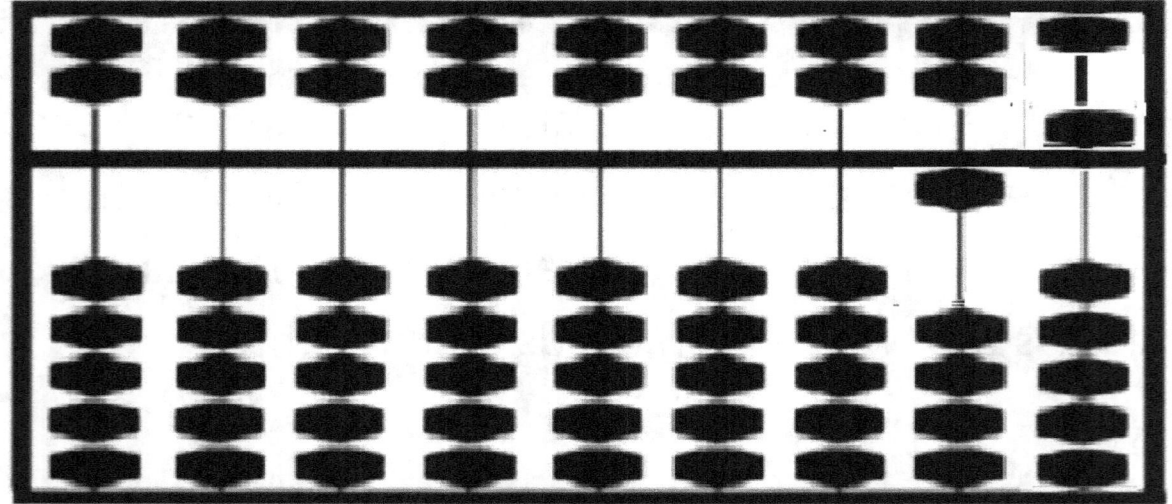

The beads touching the bar have the values of a 1 and a 5. There is, still 15 at the bar.

Whenever we have all five Earth Beads up to the bar we need to move a Heaven Bead on the same column down to the bar and move the Earth Beads back to the bottom on that column.

Addition

When we add we simply move beads toward the bar. If we need to add a number and the number is greater than the available beads can represent, then we add a greater number and subtract the excess, resulting in the addition intended.

Set 27 in the first two columns on the right. That will be 2 on the 10s column and 7 on the 1s column.

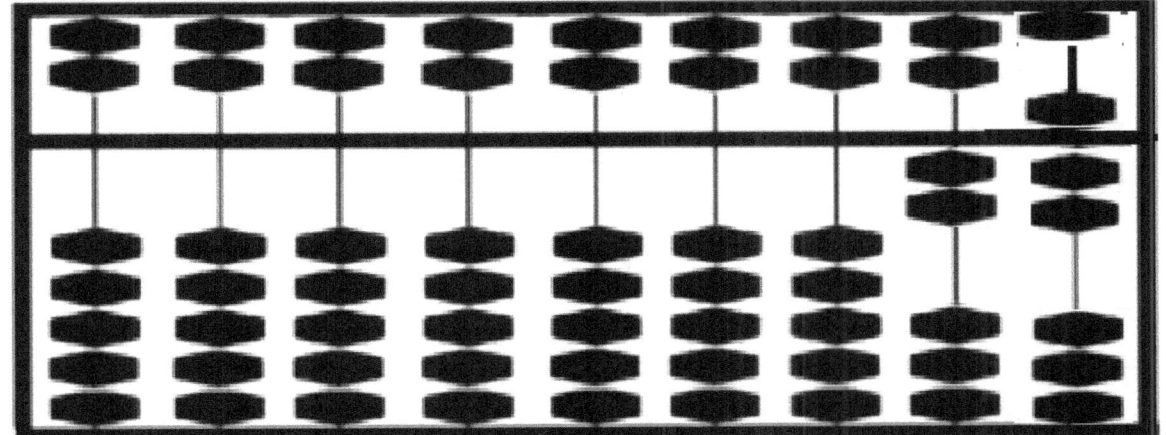

Now we will add 49 to what we have.

We need 1 Heaven Bead and 4 Earth Beads to make 9. We move a bead up on the 10s, and we move a bead down on the 1s. (+10-1=+9). We do not have 4 beads to move up for 40 on the 10s column, so we move down 1 Heaven Bead and move down 1 Earth Bead on the 10s, (+50 -10= +40).

We can read the beads at the bar as, 7 and 6.

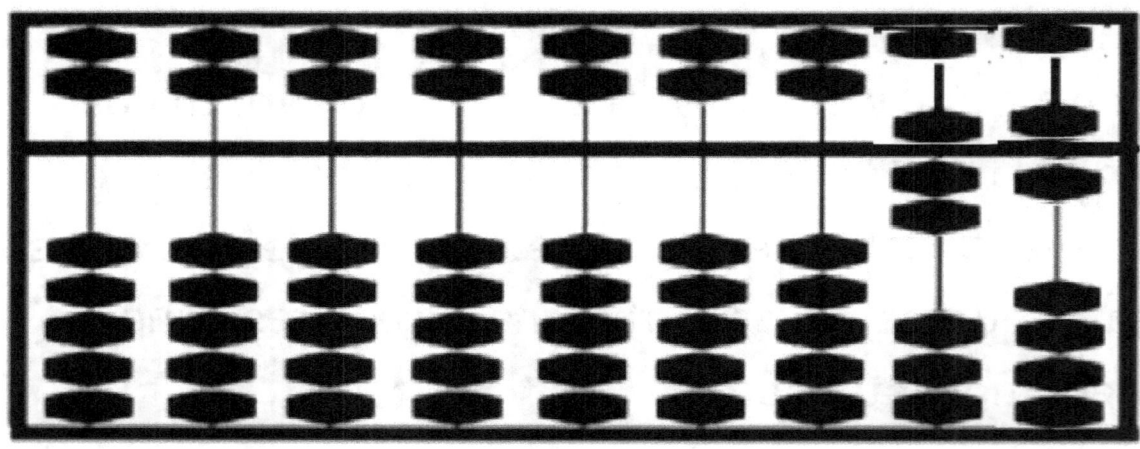

(27 + 49 = 76)

Try these on the abacus:

6+8=   12+7=   18+9=   20+14=

175+22=   321+99=   500+108=

Check your answers on an electronic calculator or by hand.

Subtraction

To subtract we move beads away from the bar.

Sometimes when subtracting a number we need to do the opposite of when adding. We may need to subtract a larger number than we intend and add back the excess!

Set 388 on the abacus

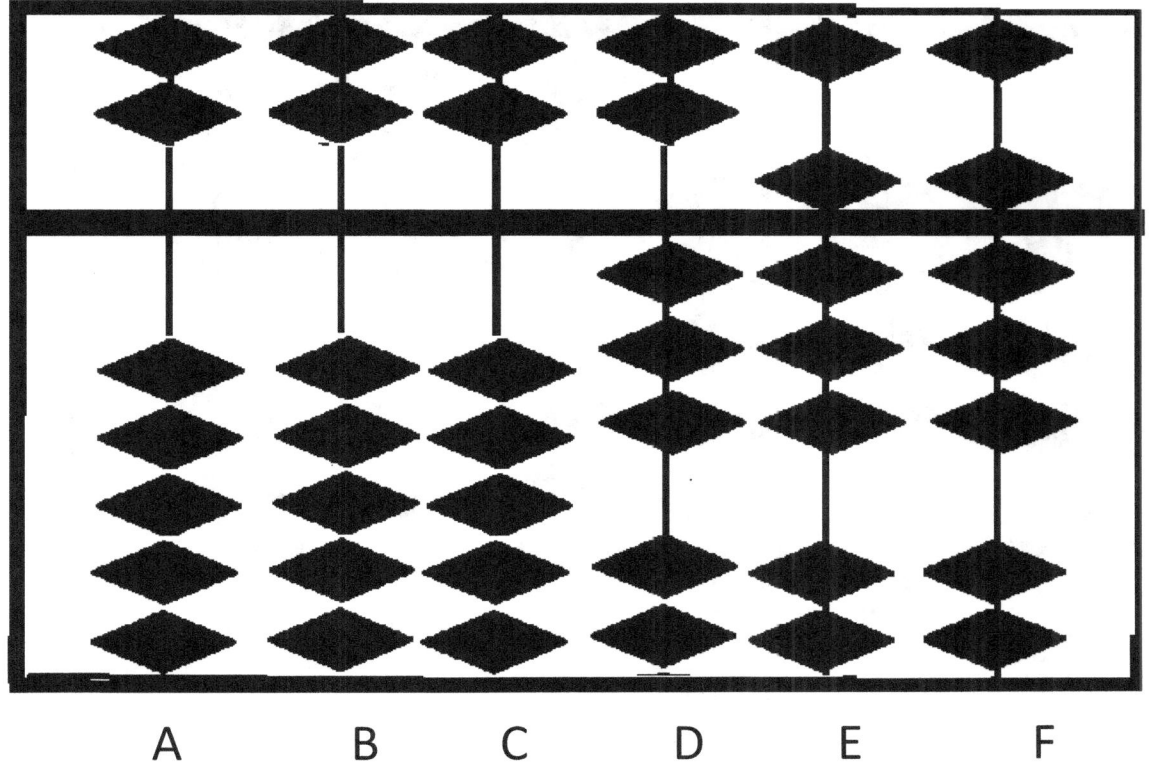

Subtract 159

First we can move an Earth Bead down on column

D (-100), then on column E we can move a Heaven
Bead up (-50), but to subtract the 9, we move an Earth
Bead down on column E (-10) and move an Earth bead
up on column F (+1). The result is 388-159 = 229.

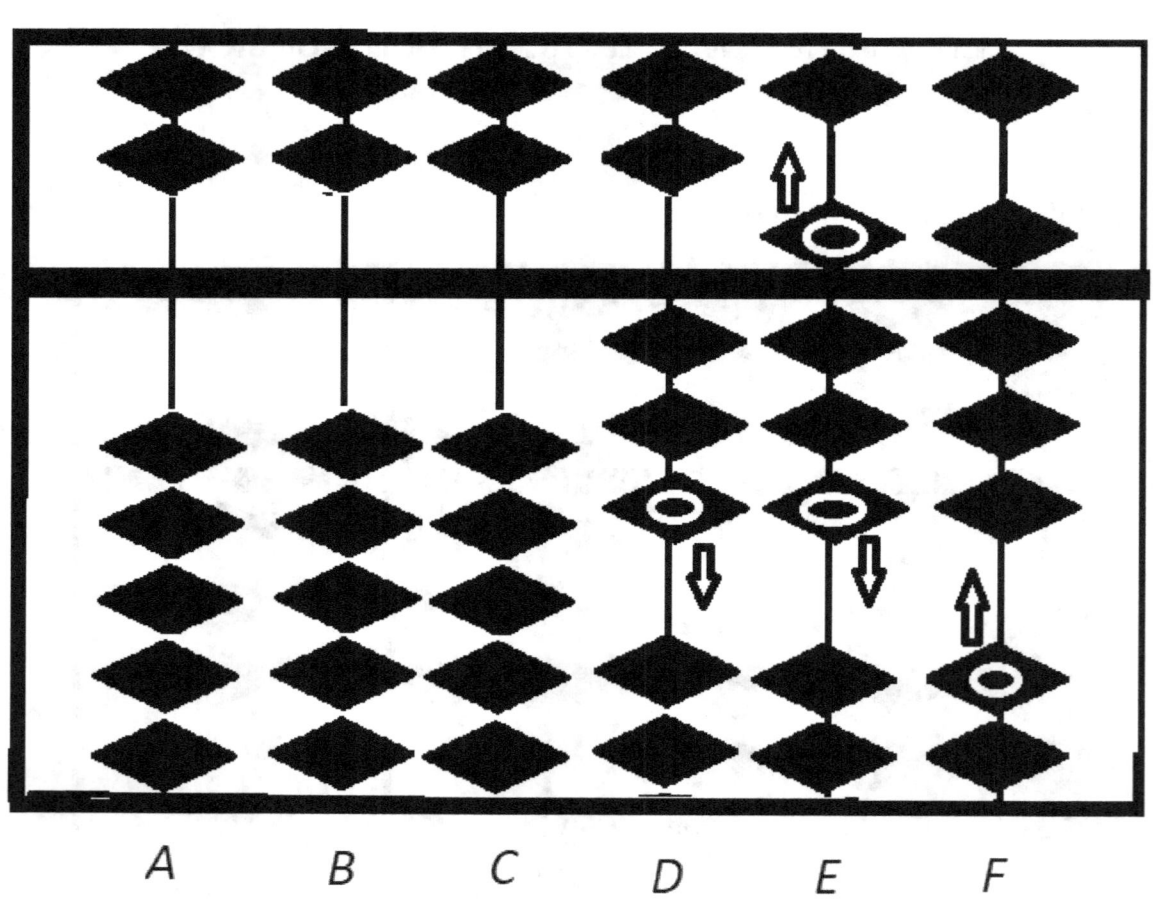

Try these now:

22-18=   110-104=   308-25=   100-99=

247-228=   155-65=   930-741=

34-25=   206-119=   75+12-42=

Multiplication

When we multiply or divide, we will have two numbers on the abacus to carry out our operations.

We want to be sure to leave at least one null column between those numbers and more if we have room, for example: multiplicand and multiplier, or divisor and dividend.

There is always a need for extra caution when working with numbers ending in zero. A column holding a 0 can be mistaken for an empty column. For numbers ending in zero, leave an extra column.

On the Chinese abacus the beads are usually all the same color, but in diagrams we try to make it easier to see which beads are moved and which are stationary. Different colors and shadings may be used for this purpose. Unused or uncounted beads may be left out of diagrams on paper.

1 digit multiplier

104 x 4

Set 104 to the left and leave at least one column as a separator between the multiplicand and the multiplier.

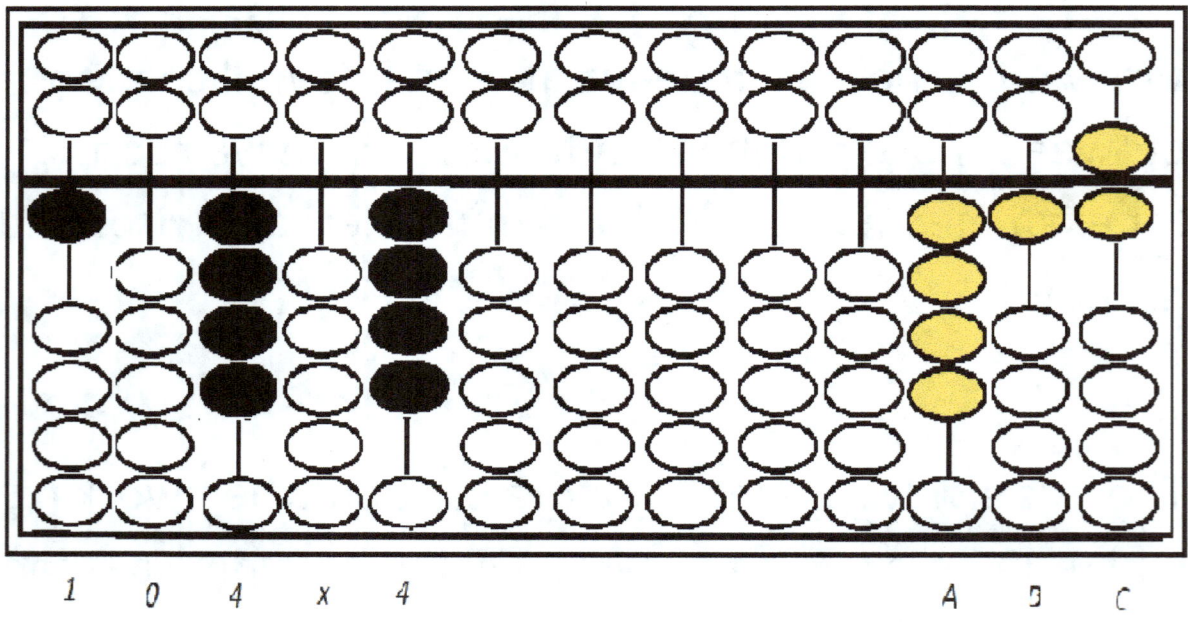

1   0   4   x   4   A   B   C

104 x 4 = 416

The answer starts with the base at column C and it is 16 (4x4); for the next digit of the multiplicand the base is at column B, and it is 0.

(4 x 0), the base then moves to A for the next digit and is 4 (4 x 1).

We started the answer at C because we did not know how many spaces (columns), we would need for the answer.

Try these:

15 x 5 =   204 x 6 =   19 x 7 =   129 x 9 =

2 digit multiplier

For the rightmost digit of the multiplier the base is at

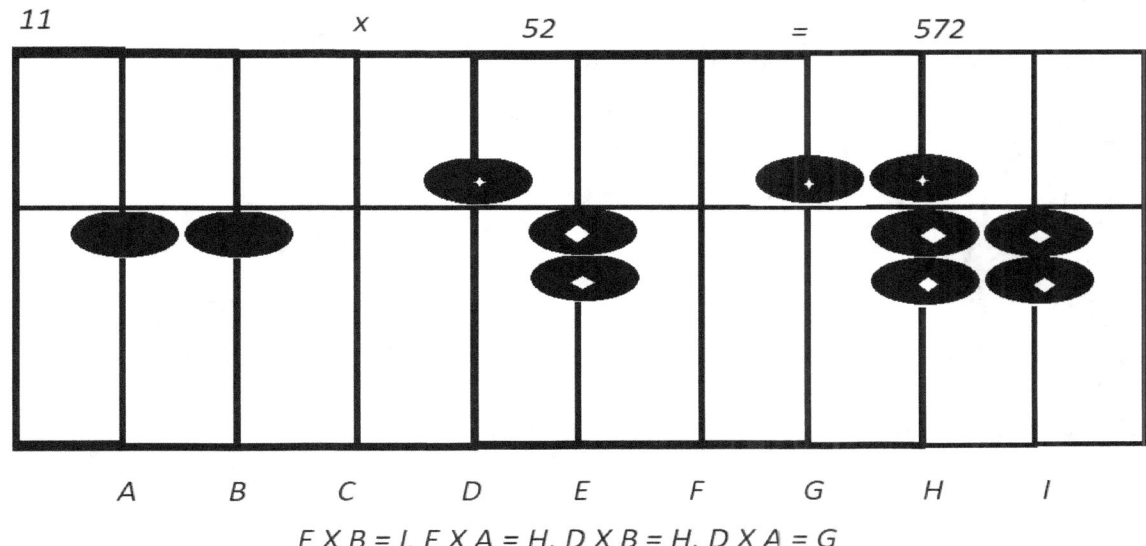

E X B = I, E X A = H, D X B = H, D X A = G

the rightmost column. As the digits of the multiplier are applied to the digits of the multiplicand, the base is moved one column to the left.

The base is also moved when we use the next digit, going right to left, in the multiplier.

Set up 213 x 18

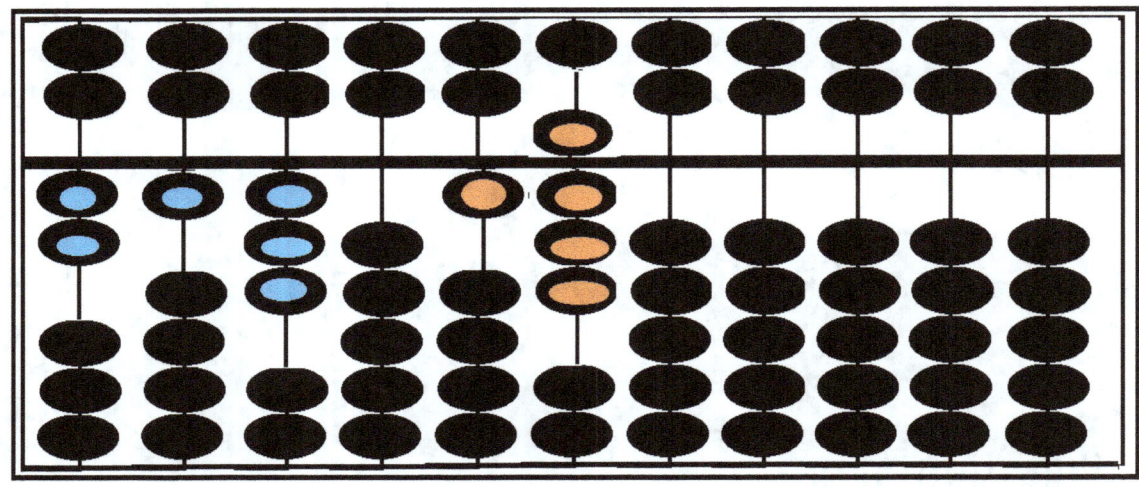

8 x 3 = 24

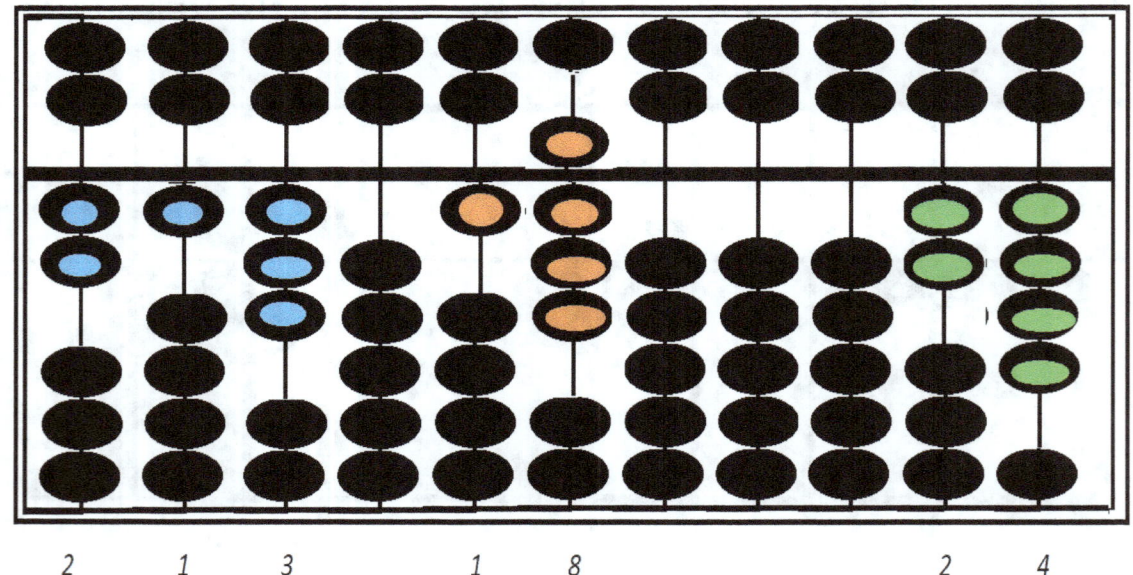

8 x1 = 8. We add 8 to what we have

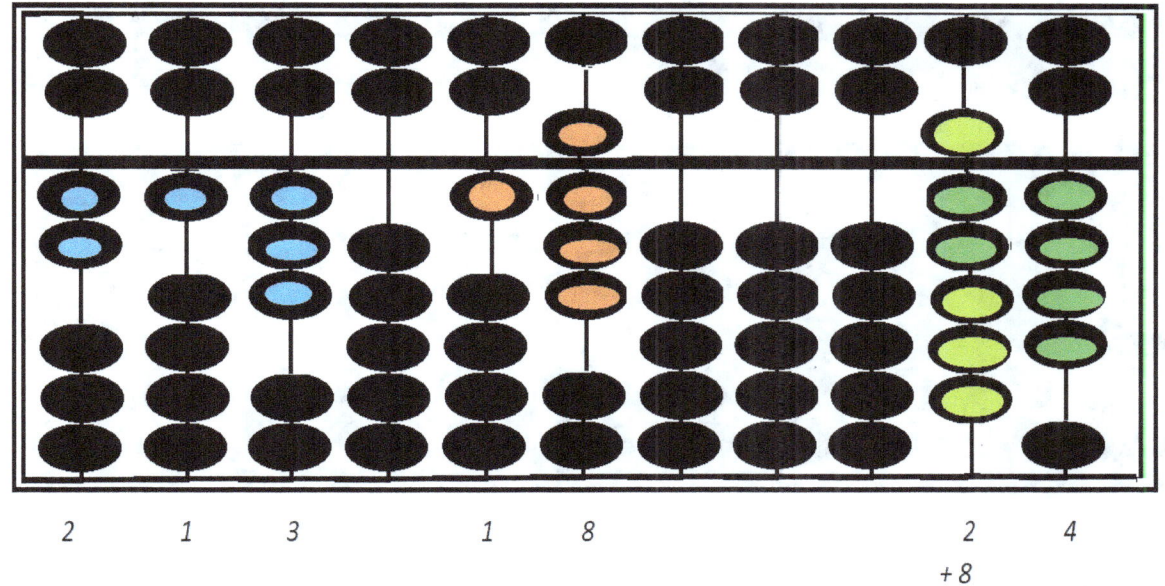

8 x 2 = 16 Added to what we have

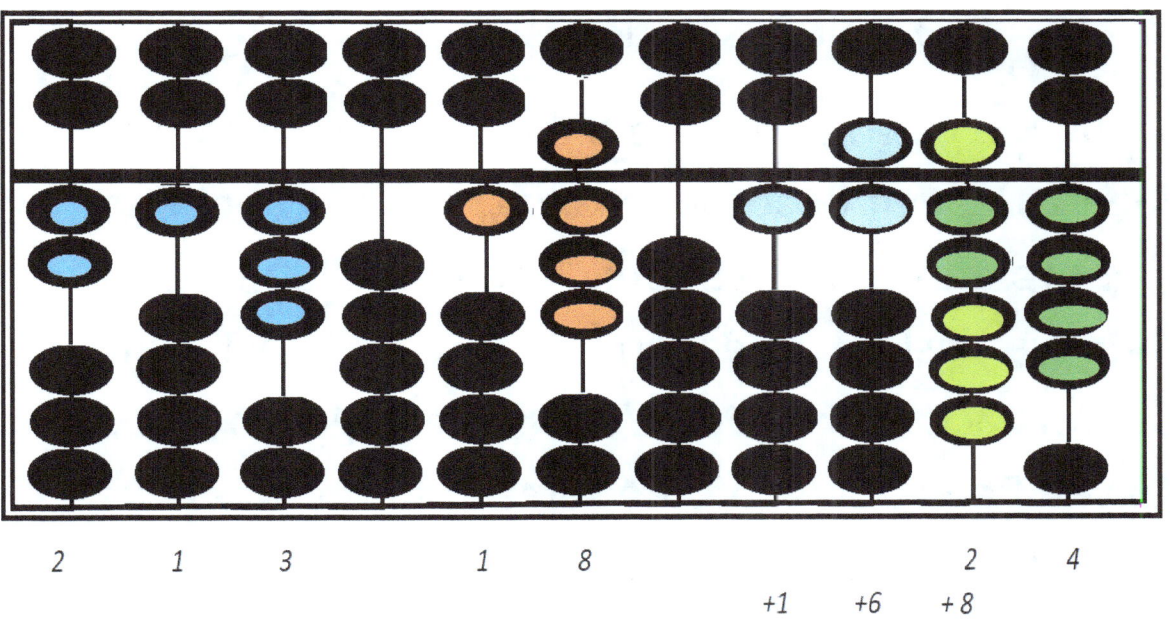

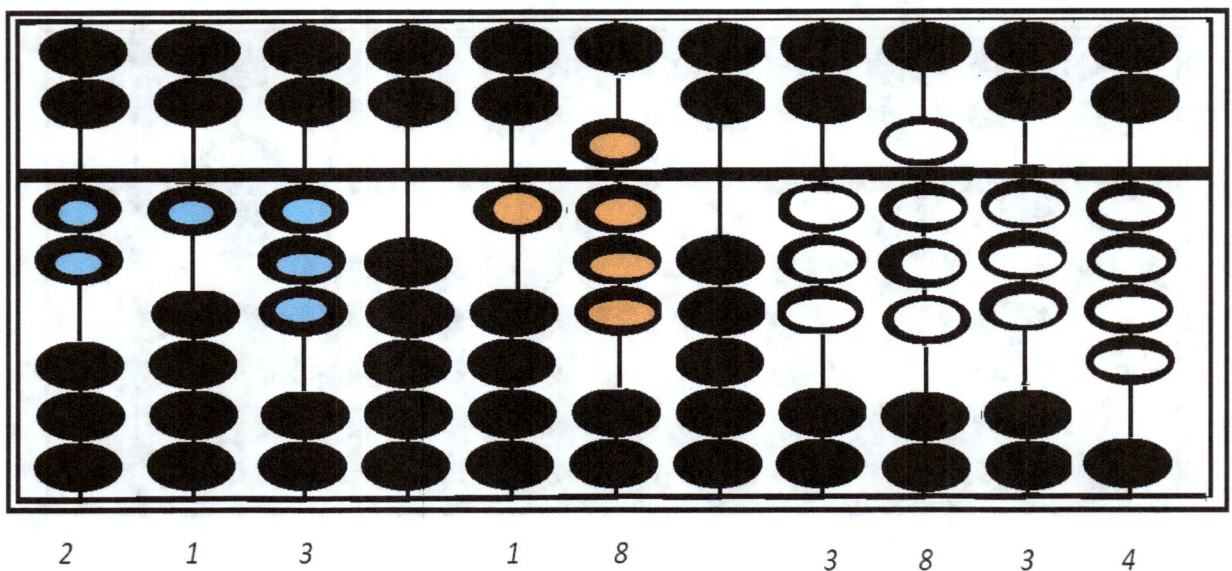

2	1	3		1	8		3	8	3	4

Now the multiplier is the 1 in 18 so the base is at the second column

1 x 213 = 213, so the solution is 3,834.

During the calculations we did not forget to adjust when 2 Heaven Beads where at the bar! We did not forget to move an Earth Bead up on the left column when we pushed 2 Heaven Beads on the next right column back to the top! We did not forget to move a Heaven Bead down, on the same column, when we moved 5 Earth Beads down If we did then maybe we have the wrong answer.

Do this one over, until you get it right, before going on.

Try these:

20 x 35 =   179 x 42 =   105 x 41 =

199 x 20 =   209 x 32 =

3 digit multiplier

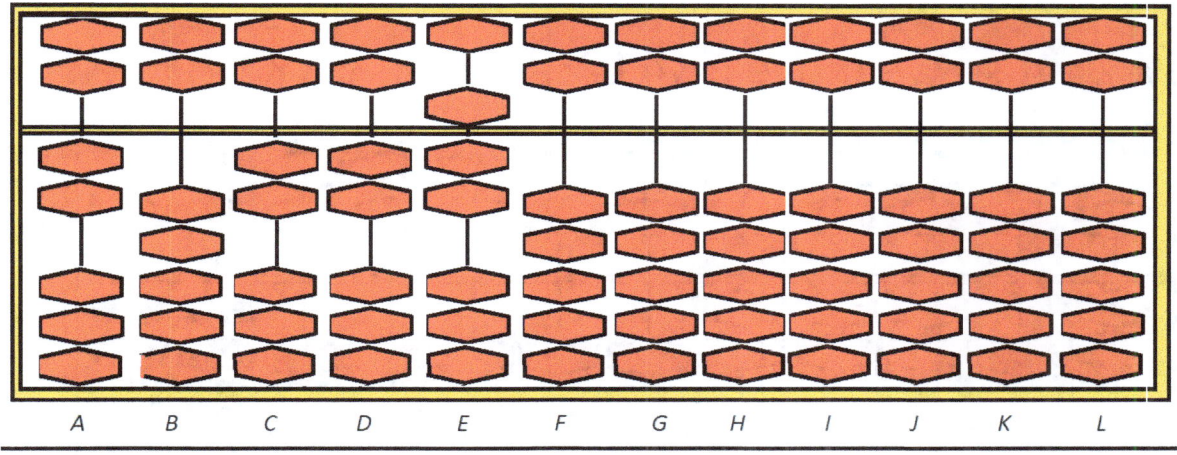

| A | B | C | D | E | F | G | H | I | J | K | L |

Let us multiply 2 x 227. The base for multiplier 7 on E is at L

For 2 at D the base is at K and for 2 at C it is J.

7 x 2 = 14 and we set 14 on the right two columns as 1 and 4

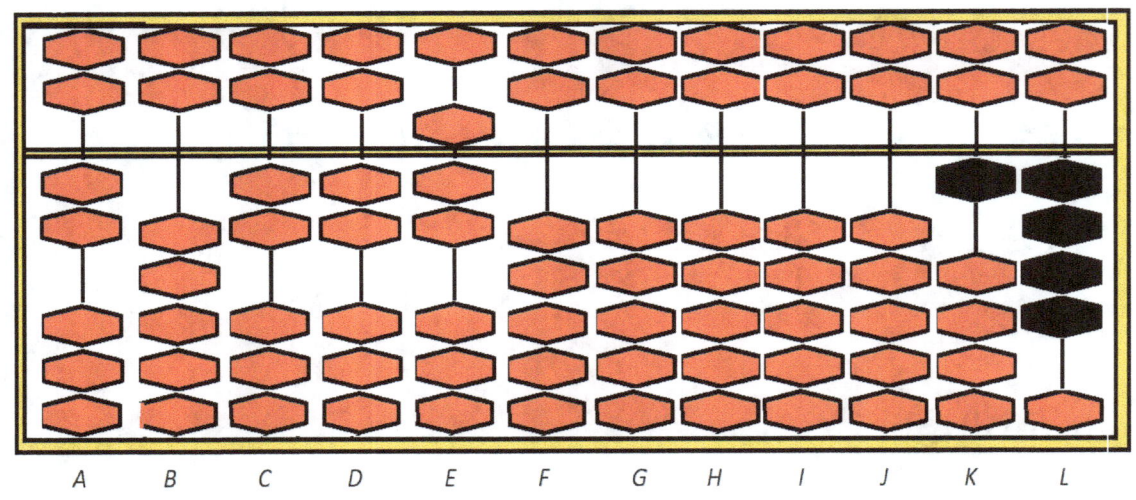

2 x 2 = 4 so we move up 4 Earth Beads on K

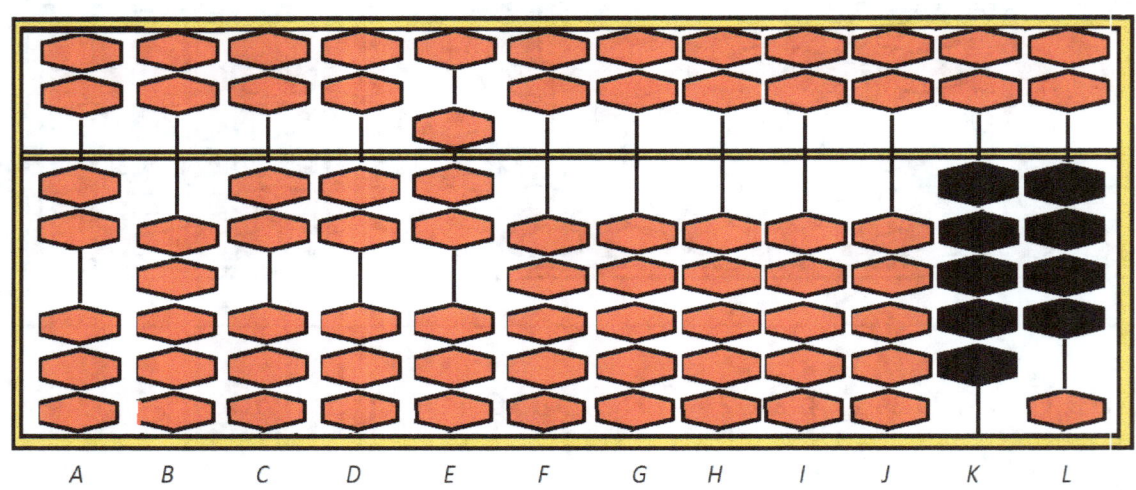

2 x 2 = 4 so now we add 4 Earth Beads on J

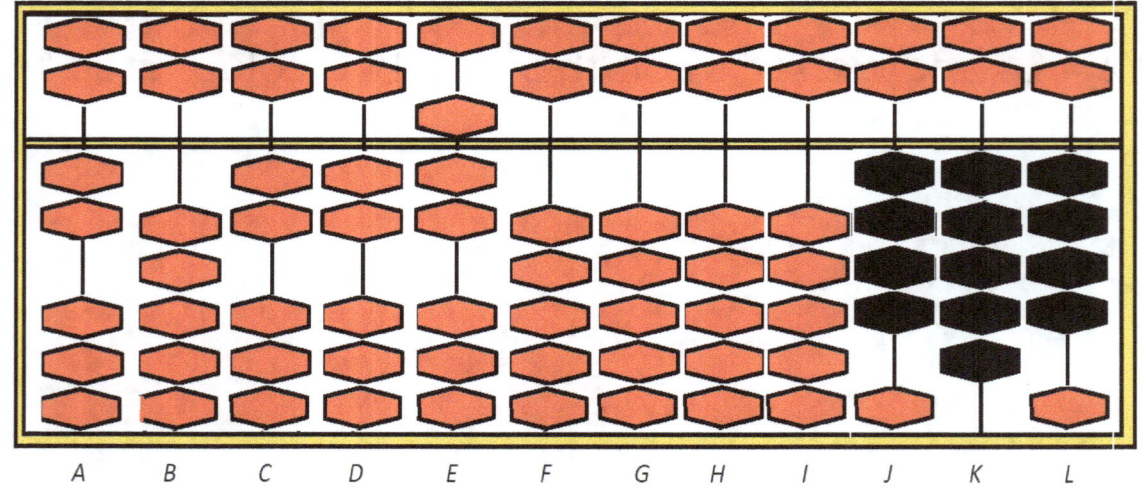

Now we need to adjust for K, all Earth Beads are used

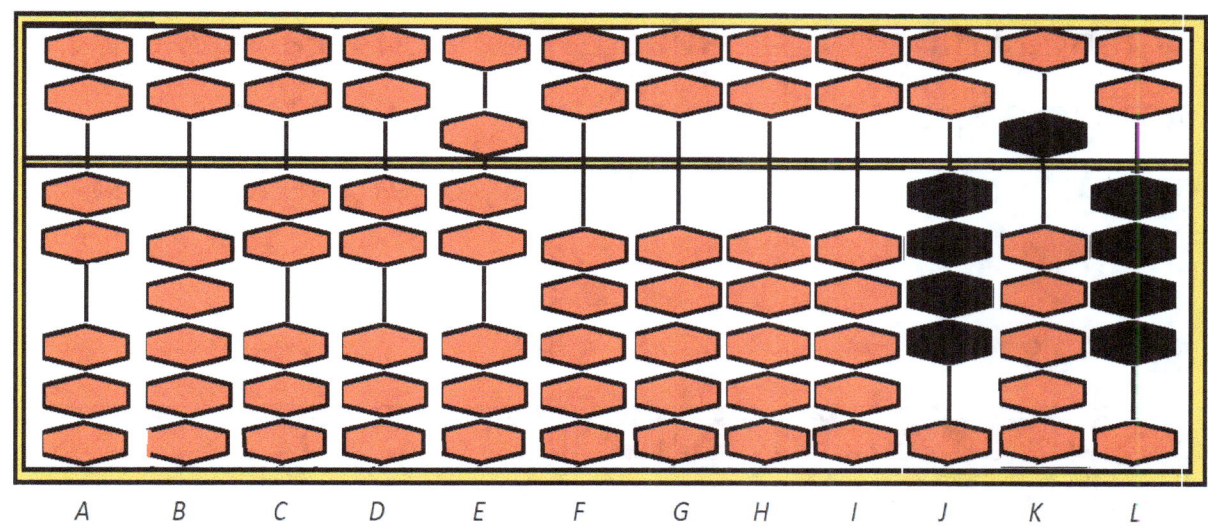

and they are replaced by 1 Heaven Bead, we have

2 x 227 = 454.

Try these..

4 x 129 =	6 x 206 =	109 x 342 =
240 x 120 =	22 x 123 =	118 x 246 =
992 x 183 =		

Division

When dividing on the abacus we need to allow sufficient spaces for the quotient, so we start our answer entry close to the divisor, or at the leftmost column.

414 / 18

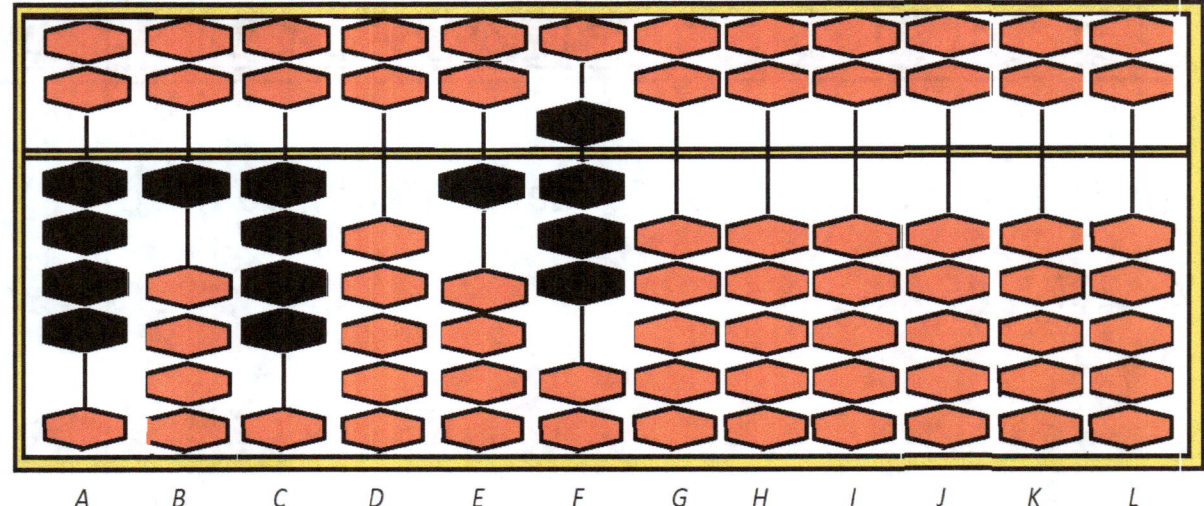

18 goes into 41 twice (2 x 18 = 36), 41 − 36 = 5

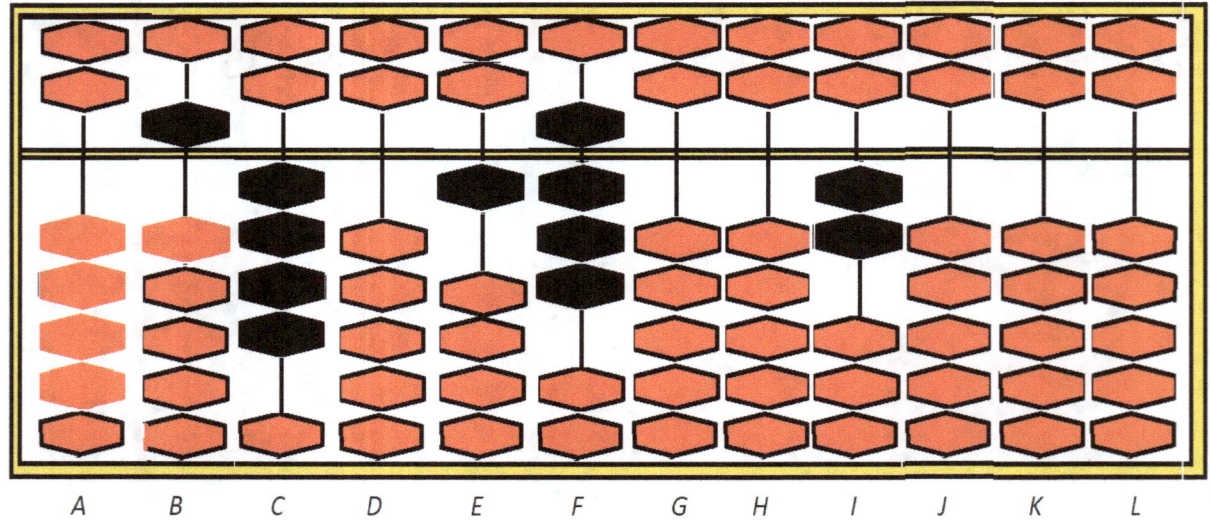

We have reduced the 41 by 36 as you can see on the abacus we now have 54 to divide by 18. 18 goes into 54 three times (3 x 18 = 54).

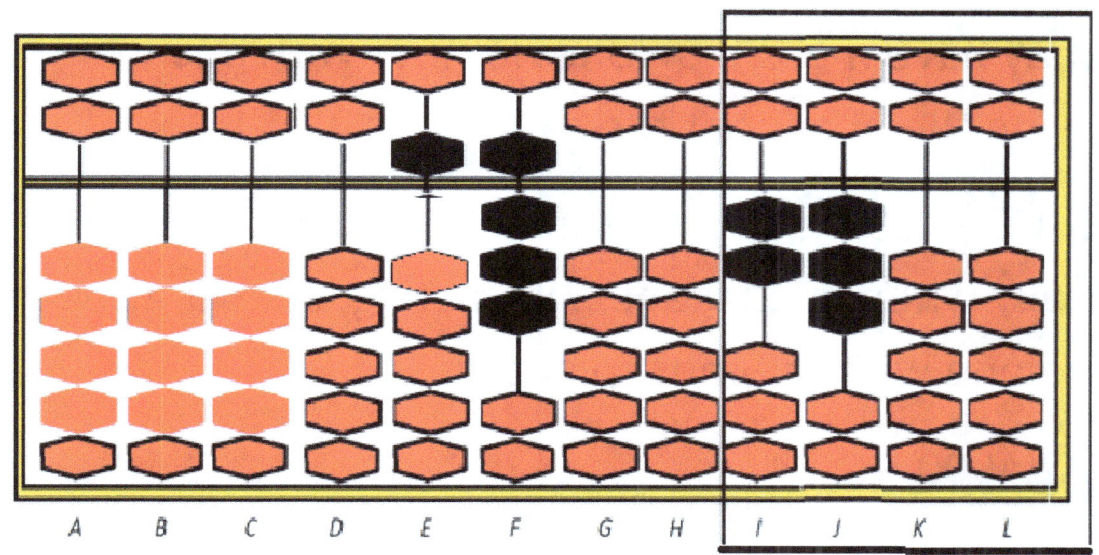

Add the 3 Earth Beads to J. Subtracting 54 leaves us no remainder with 23 as the quotient.

Can you complete these?

225 / 25 =   216 / 3 =   608 / 19 =

135 / 15 =   658 / 72 =

Decimals

On the abacus the decimal point is wherever you designate.

You may have more than one number with decimal points on the abacus at the same time.

32.45 + 3.1416 (keep all decimal points lined up vertically when adding and subtracting, allowing space for the one with the most places.)

```
    32.4500

+   03.1416

    35.5916
```

We need to plan for at least 4 decimal places!

Set 32.4500

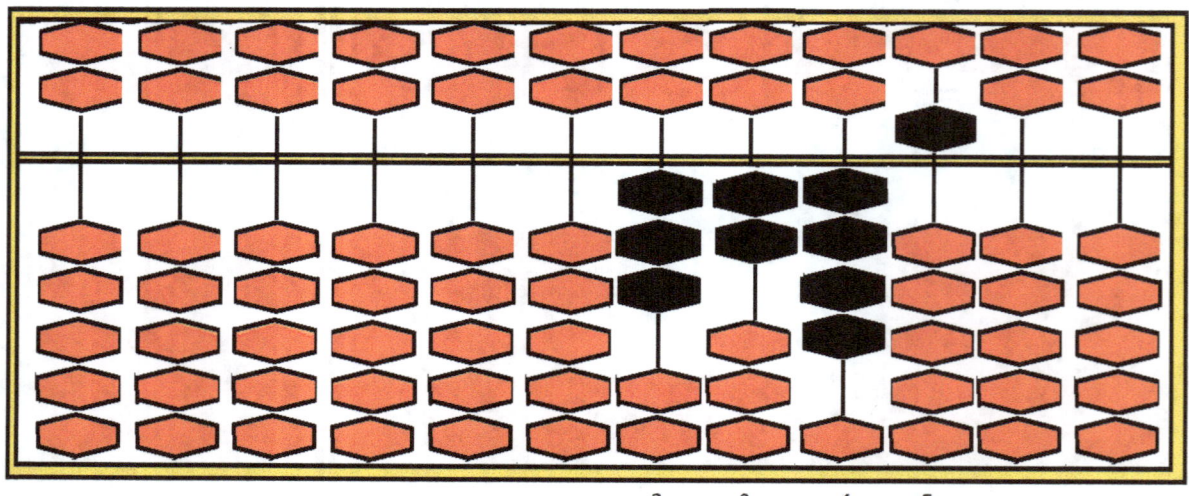

3　　2　.　4　　5

+ 3.1416

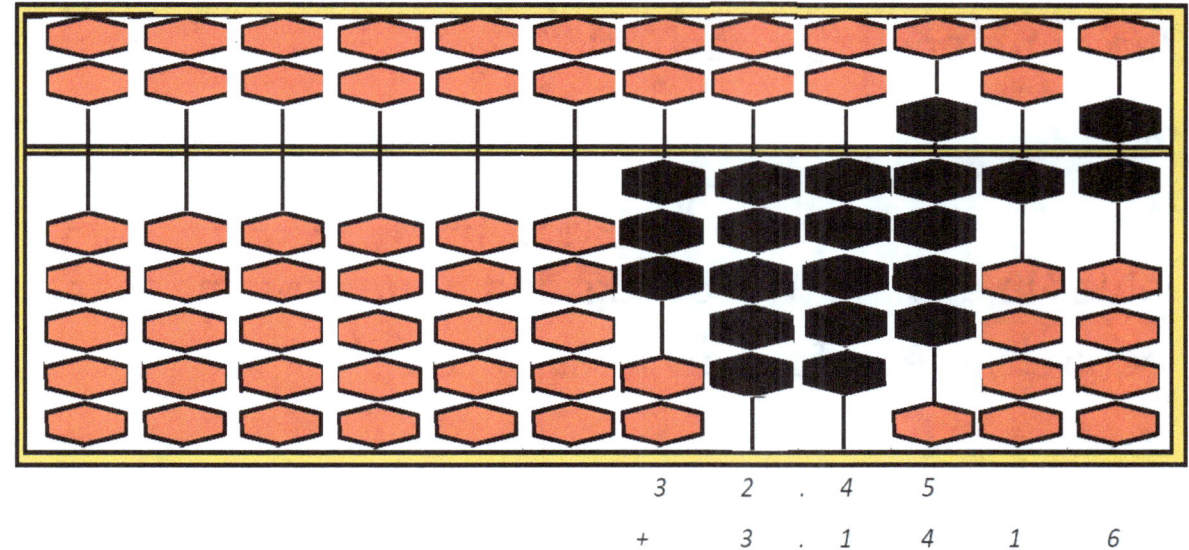

	3	2	.	4	5		
+		3	.	1	4	1	6

Replacing 5 Earth Beads with 1 Heaven Bead: then

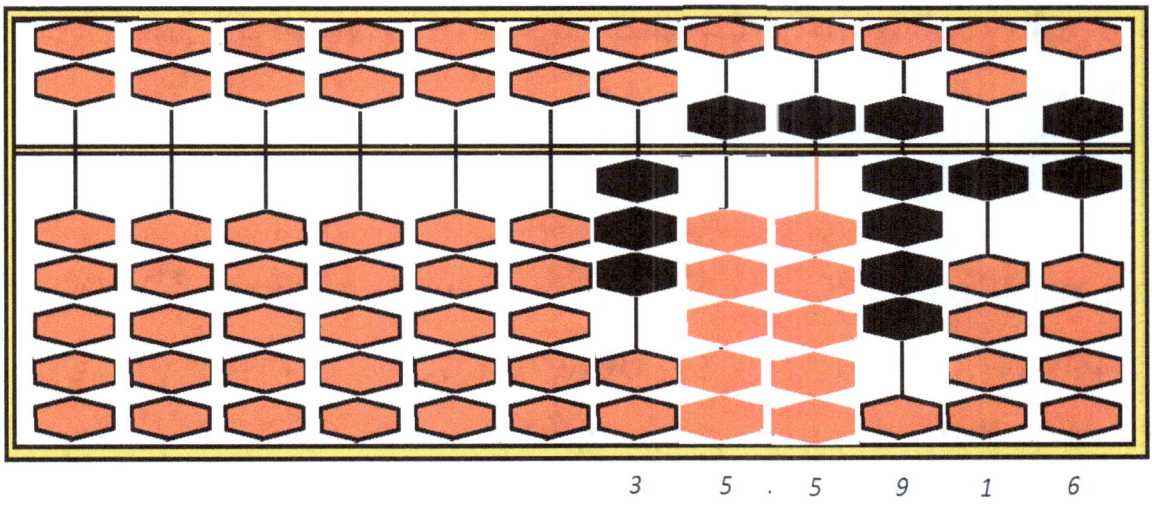

| | 3 | 5 | . | 5 | 9 | 1 | 6 |

read 35.5916

Percents

Percent means "for each 100".

Most financial calculations referring to US dollars require no more than 2 decimal places. There are some values in dealing with stocks and bonds trading that require three or four decimal places.

When a calculation result contains a fraction or a remainder, we need to convert it to the nearest decimal equivalent in order to work with it on the abacus.

A number or fraction of a number is considered 100% and for whole numbers a decimal point on the left, of the right 2 digits, gives us 1%.

Fractions are converted by dividing 100 by the denominator and multiplying the numerator by that factor.

2/25, 100/25 = 4, 4 x 2 = 8: 2/25 = 8% or .08

What is 5/20 x 6/24? .25 x .25 = .0625 (total decimal places in factors is 4 so we allow for 4 places in the answer)

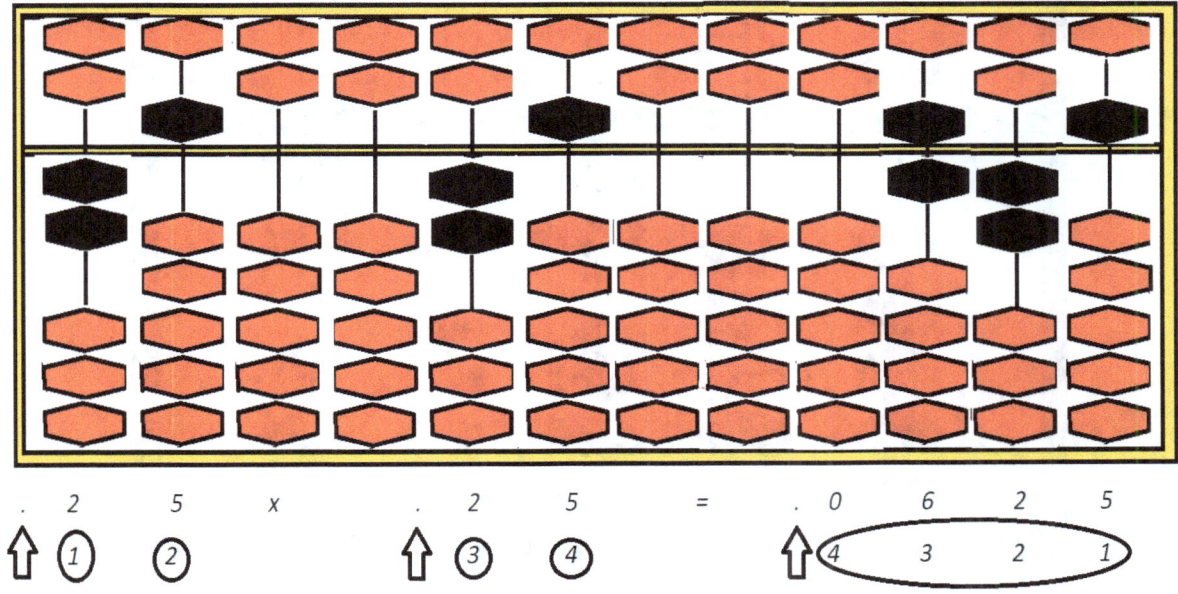

Please total these:

100-23 + 840- 19 -8 + 223- 15 +230 -102 =

(12/3 + 22) x ((6 + 14) x 15))-195 + (234/ 468) =

There are at this time several Abacus tests and certifications available online. There is no ultimate authority for certifying Abacus expertise in the United States.

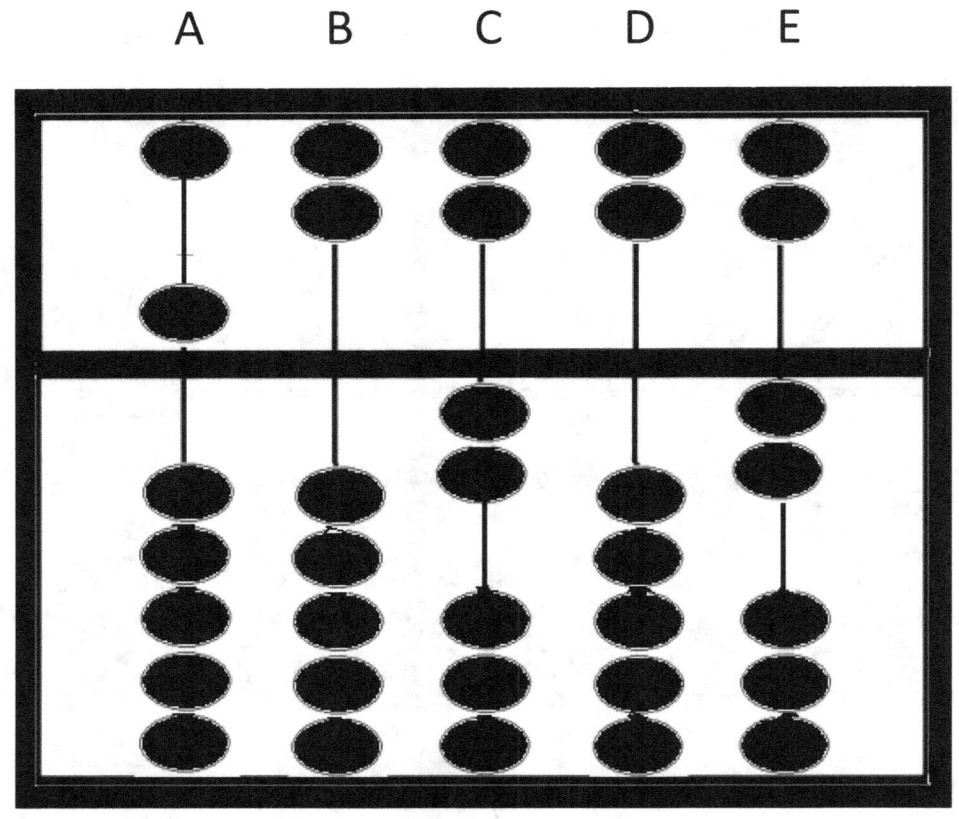

The Decimal point is between C and D

What do you read this number to be?

10202_____   102.02_____   1.202_____   .1202_____

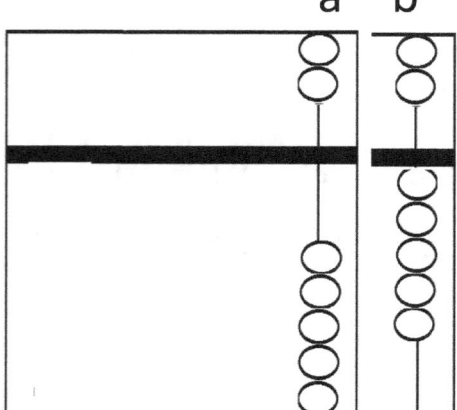

a   b   What do we do next to get from b to a?

Move a Heaven Bead down and all Earth Beads Down_____

Move only Earth Beads_____

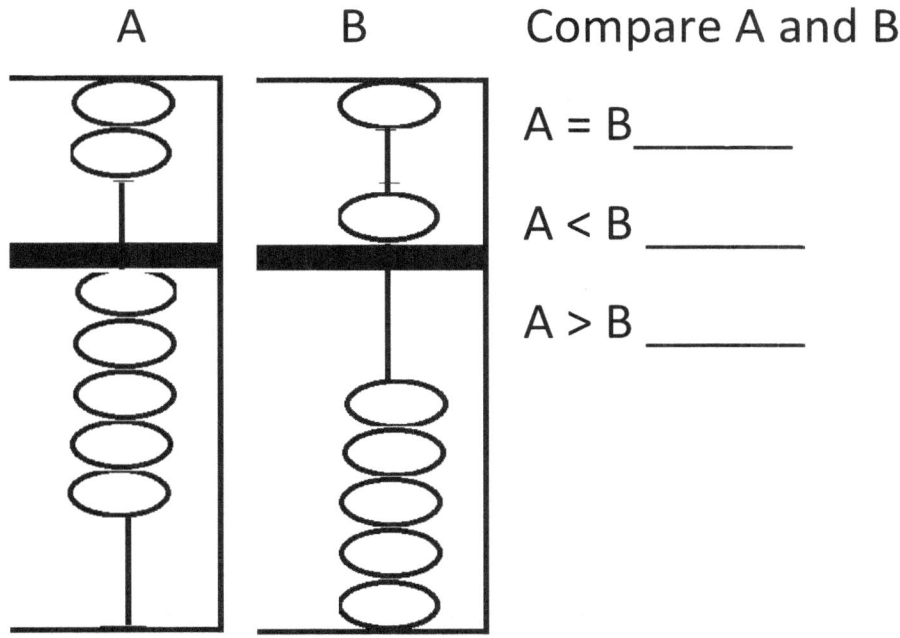

A   B   Compare A and B

A = B _____

A < B _____

A > B _____

Which comes first and gets adjusted to the other?

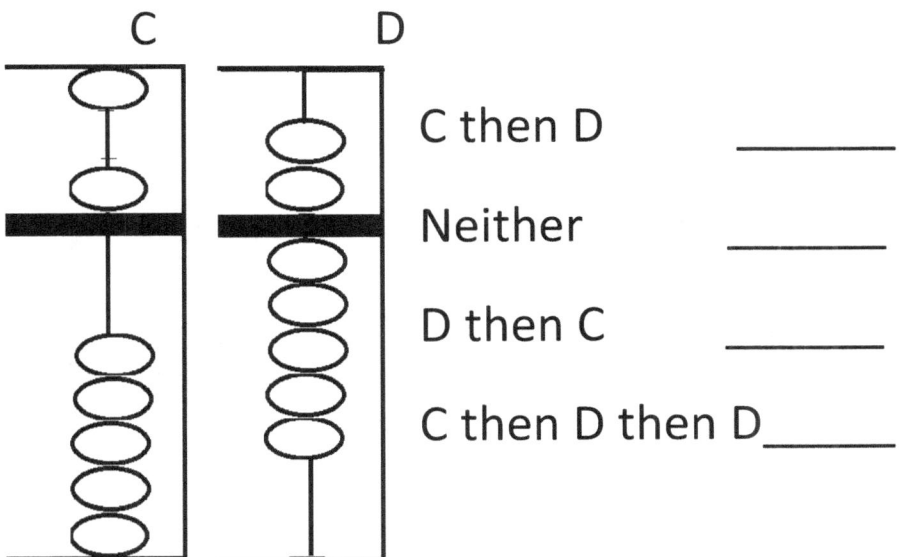

C   D

C then D   _____

Neither   _____

D then C   _____

C then D then D _____

The following is a chance for you to test yourself and determine your level of Chinese Abacus ability.

Answers will be at the end of this book and should not be read until all parts of the tests are completed.

Test A

Draw the problem and the answer beads positions.

a. 22 + 34

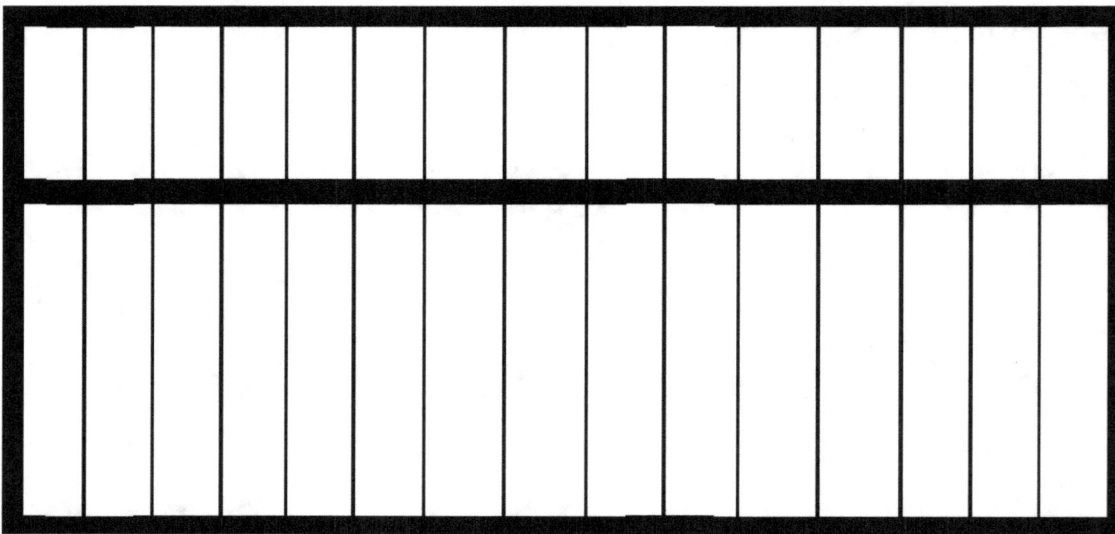

b. 16+26

c. 104+220

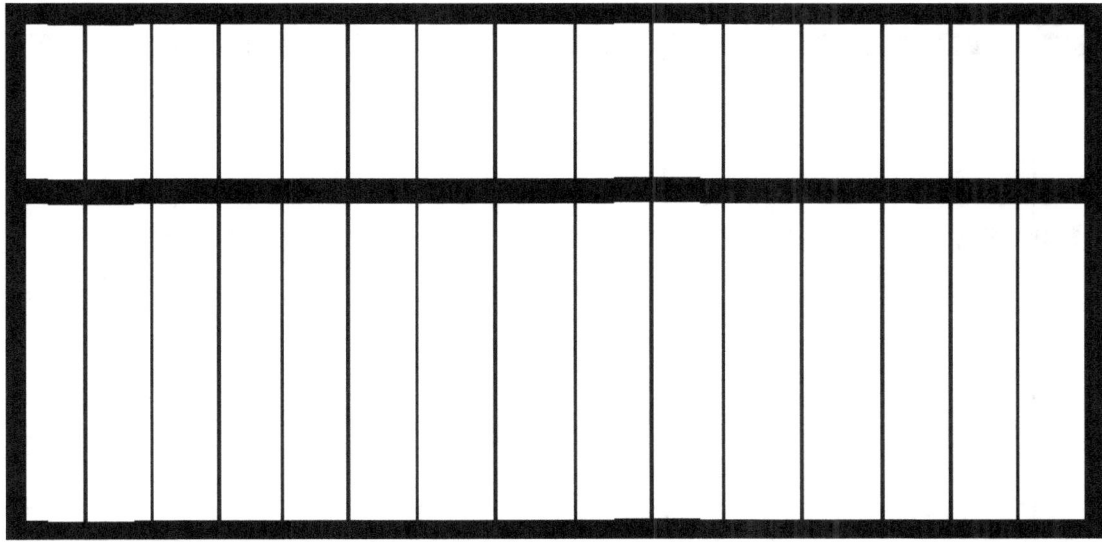

d. 30+70

e. 4004+2030

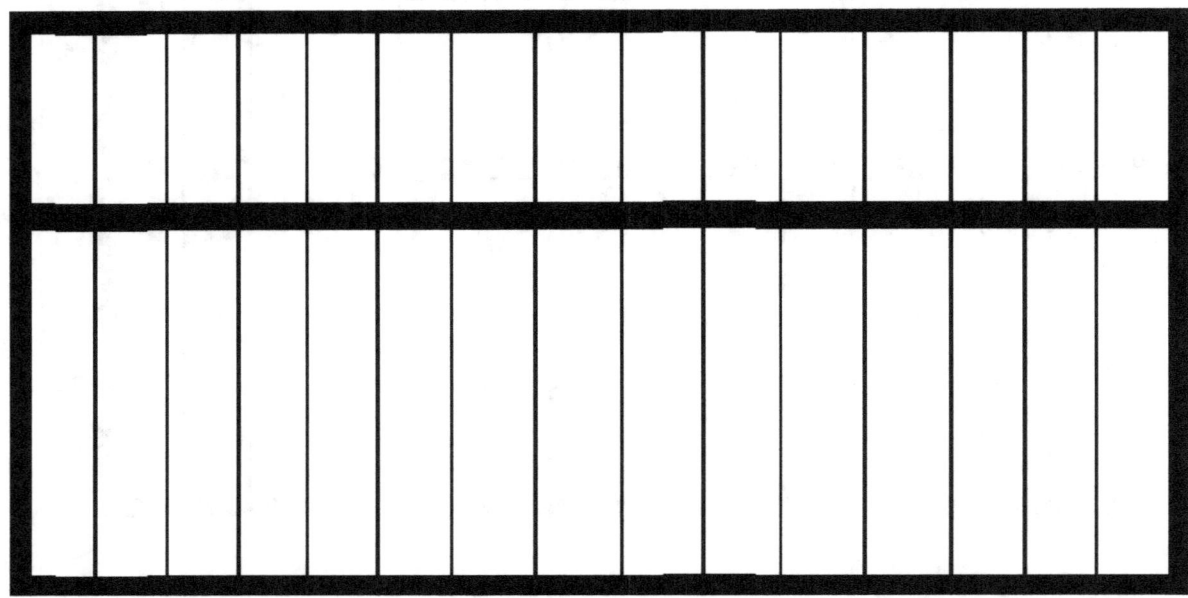

f. 180+810

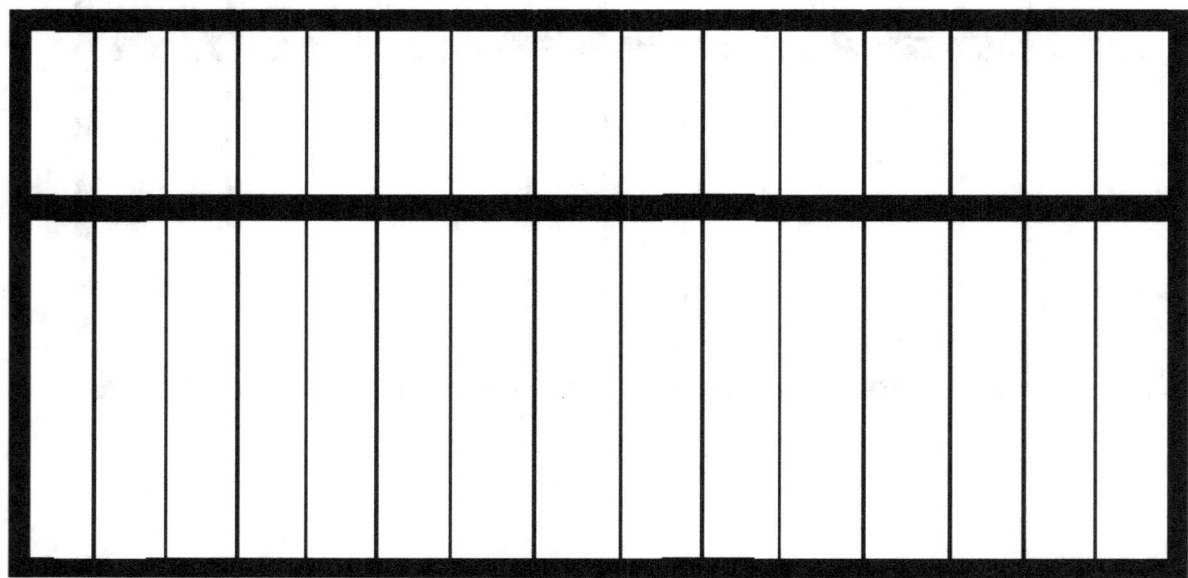

g. 999+499

h. 115+15

i. 203+219

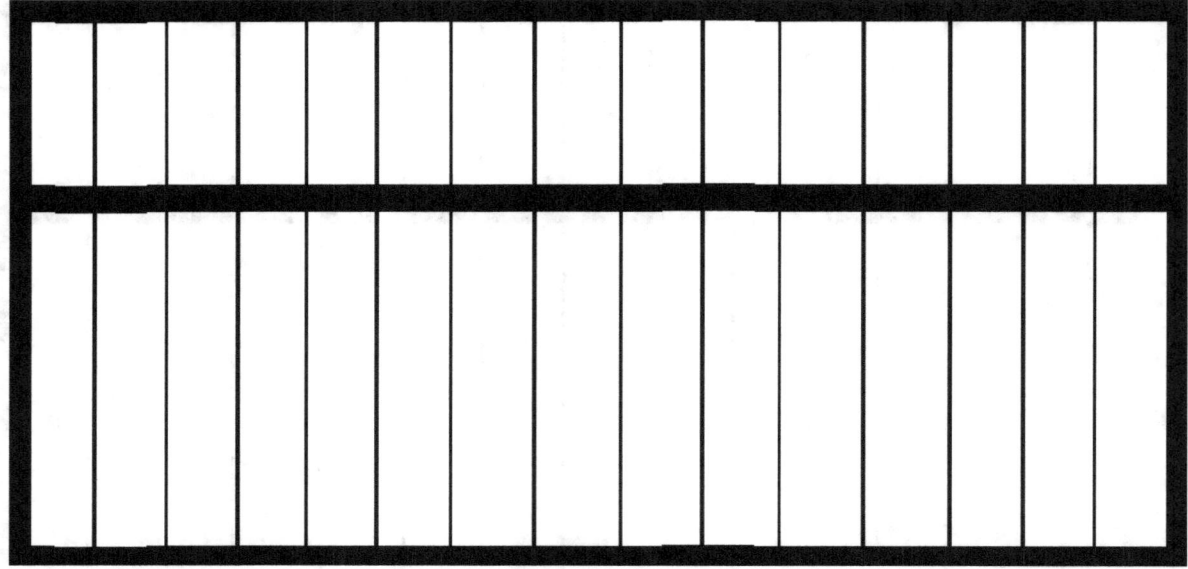

j. 4009+9090

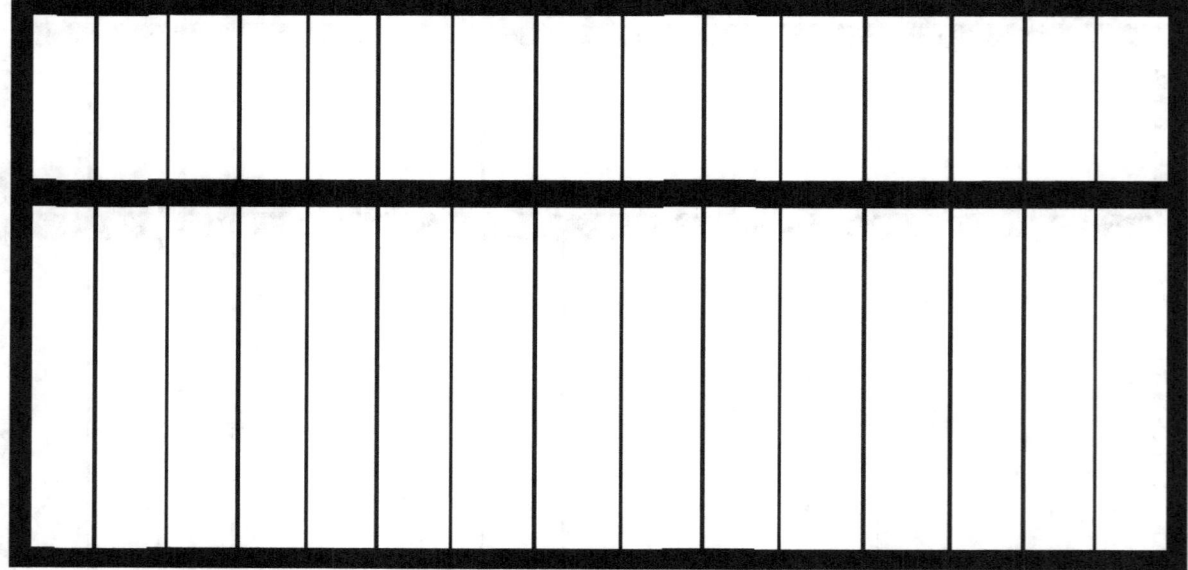

Test B, Diagram beads of minuend and result

a. 35-18

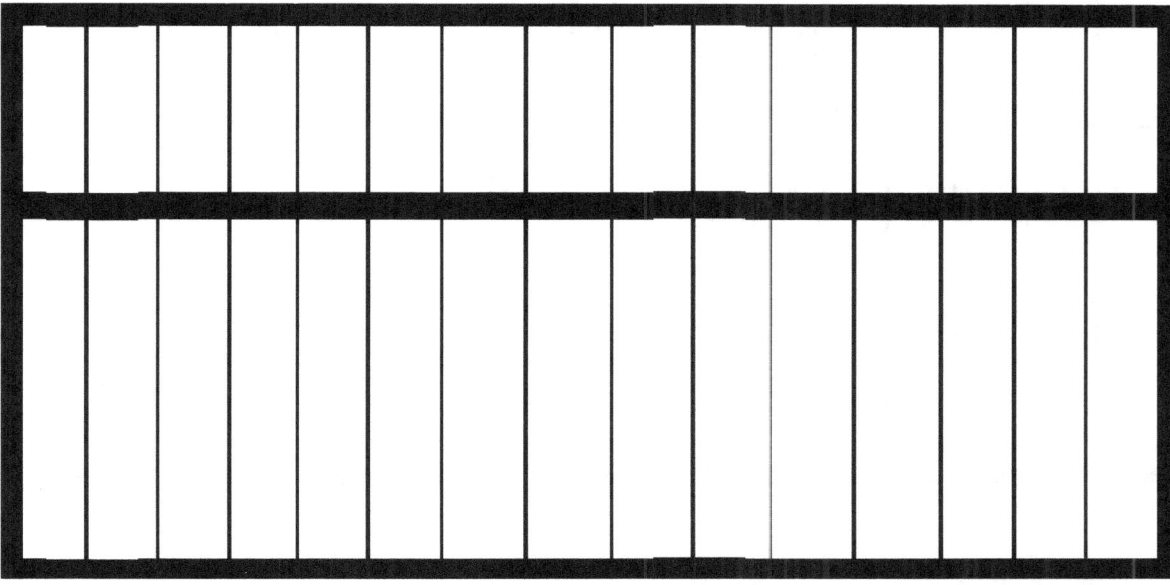

b. 355-28

c. 298-120

d. 507-203

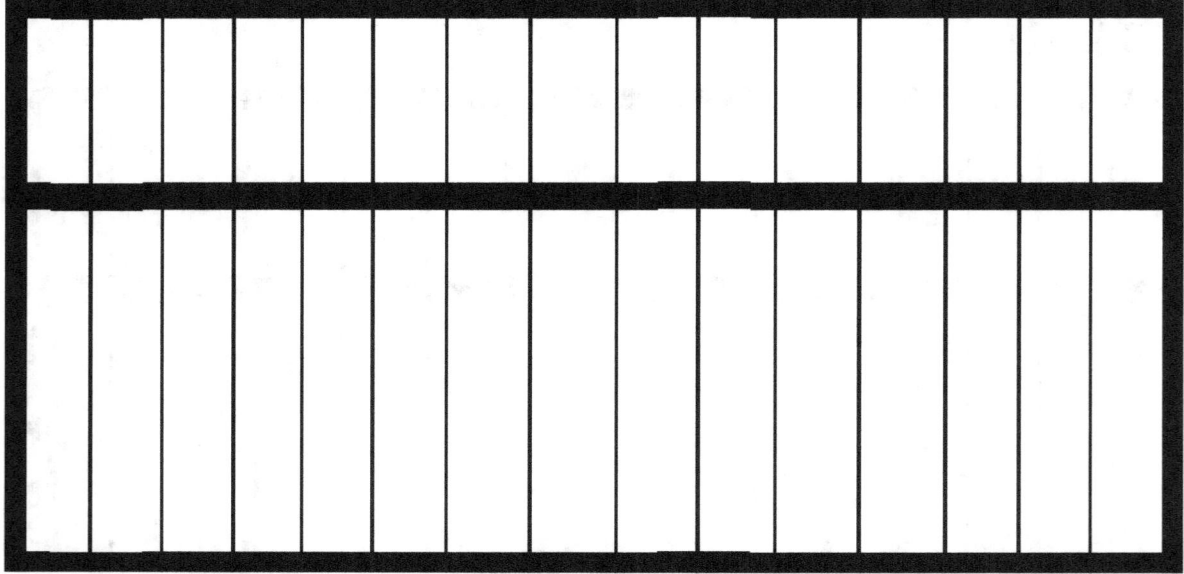

e. 3001-2003

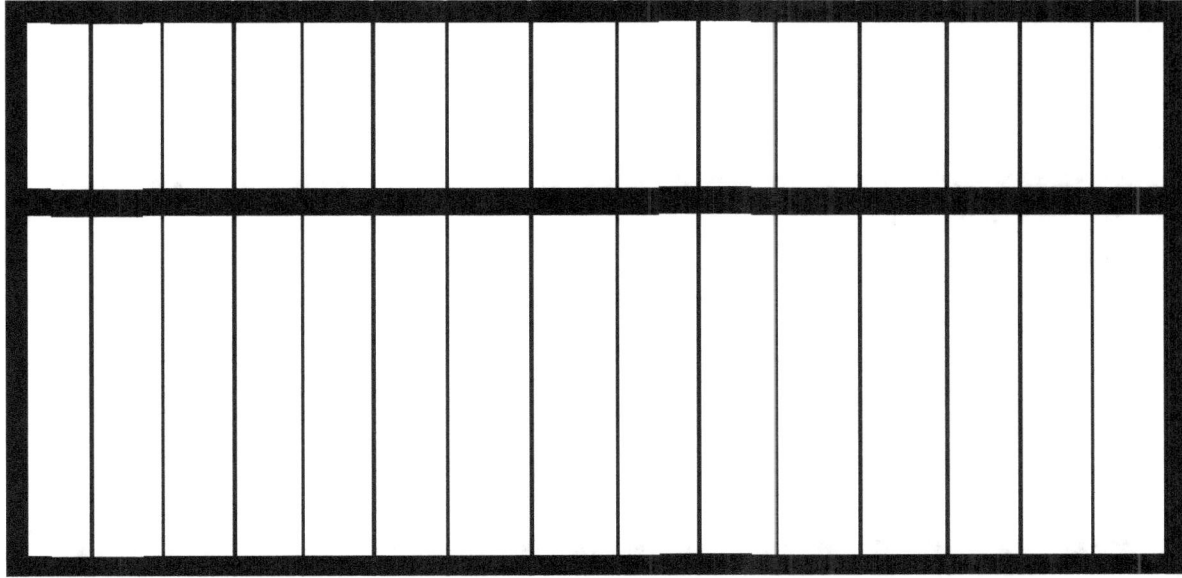

f. 100-97

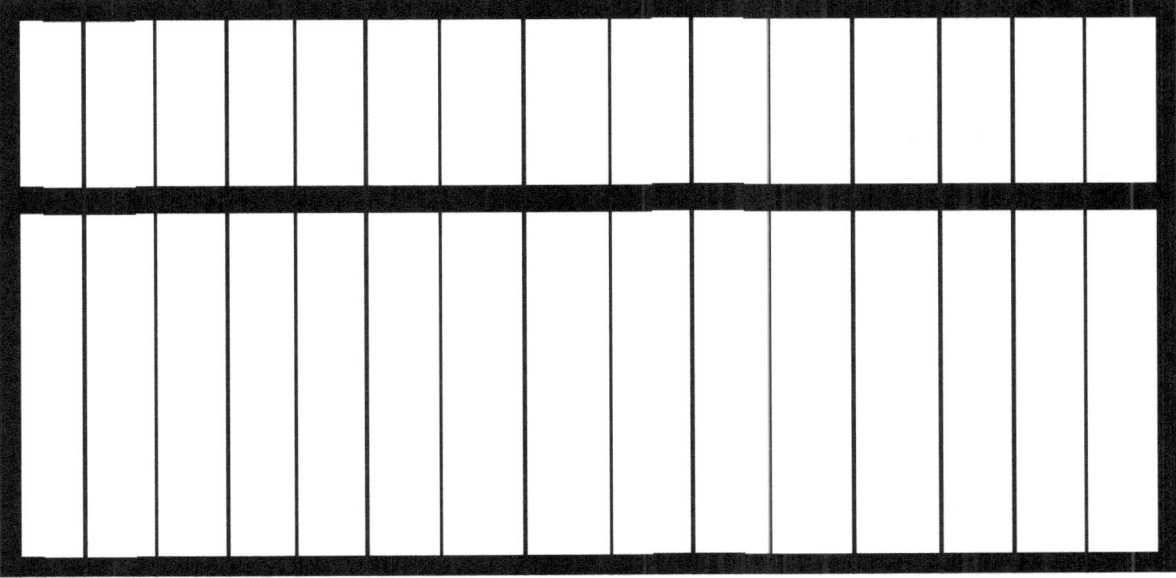

g. 222-199

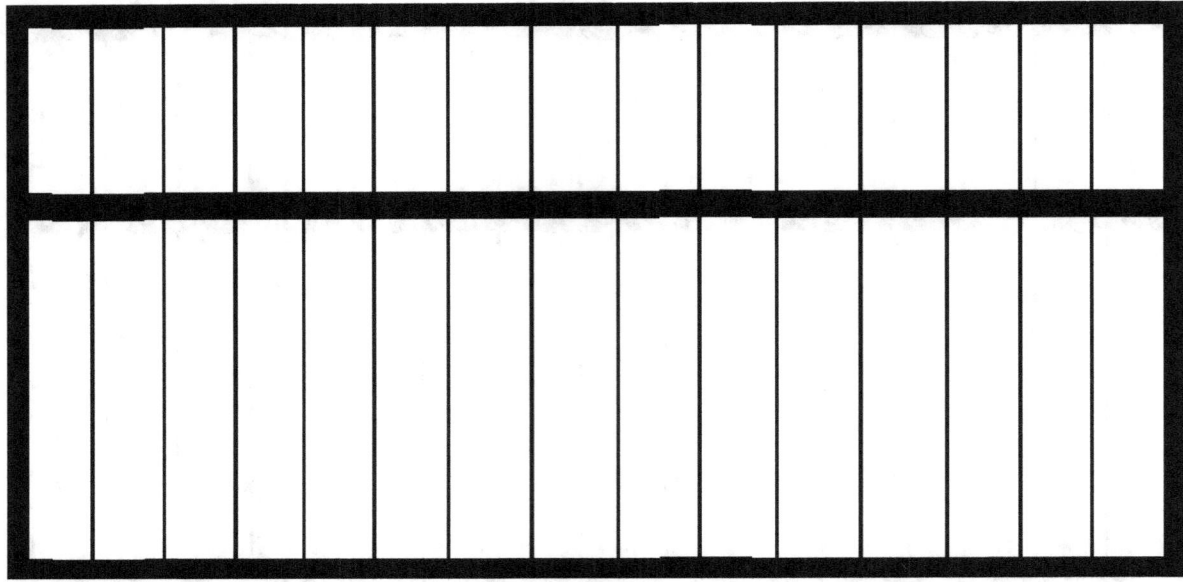

h. 606-284

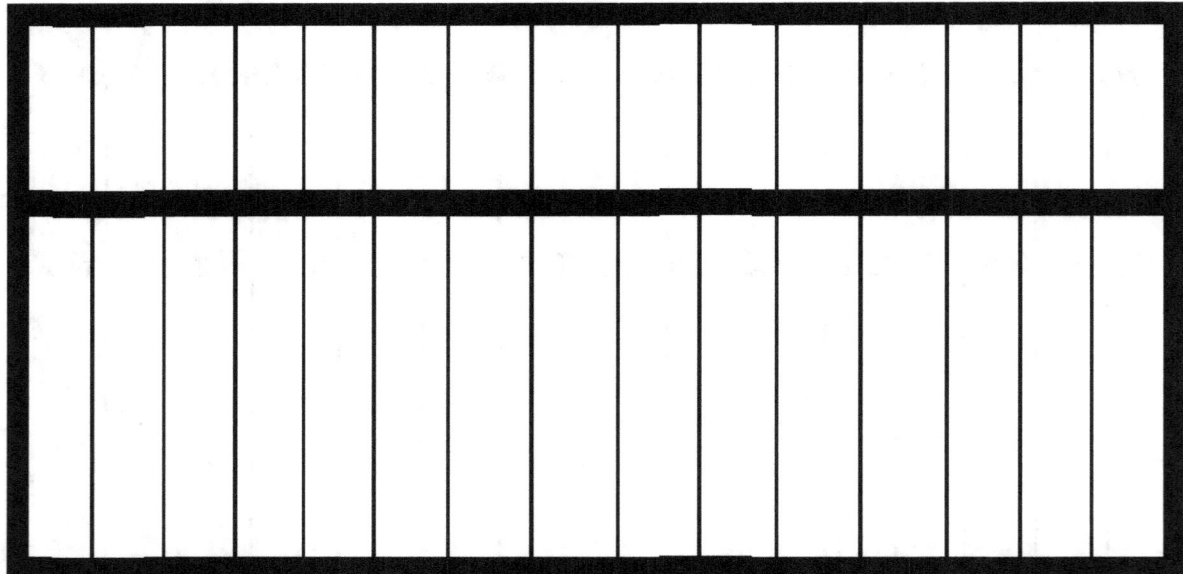

i. 903-605

j. 505-209

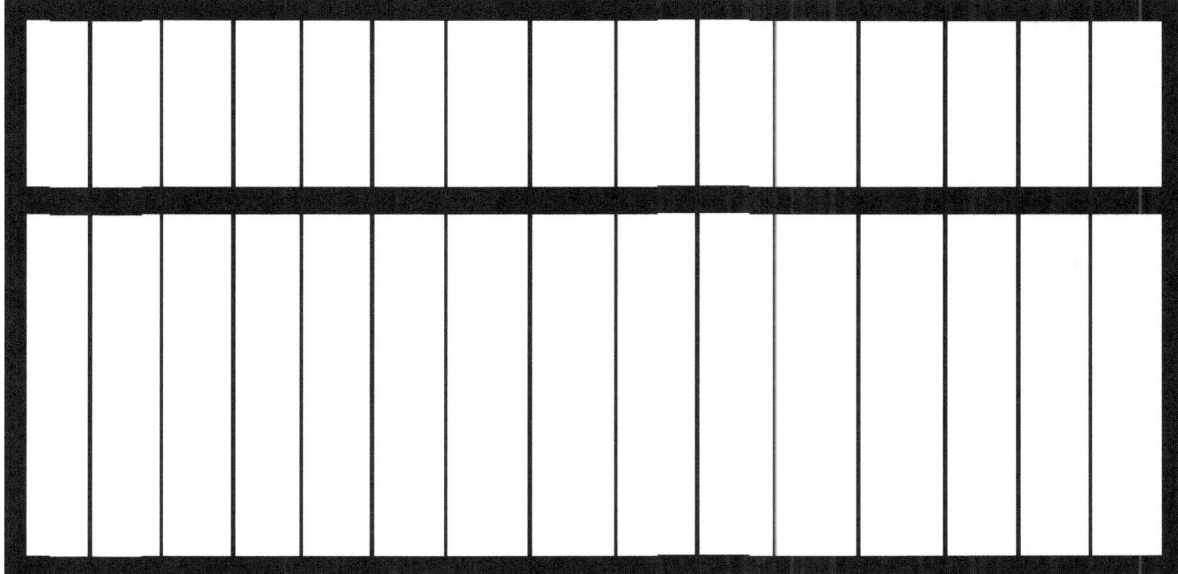

Test C, Diagram only the solution calculated on abacus

a. 23+18-9+17

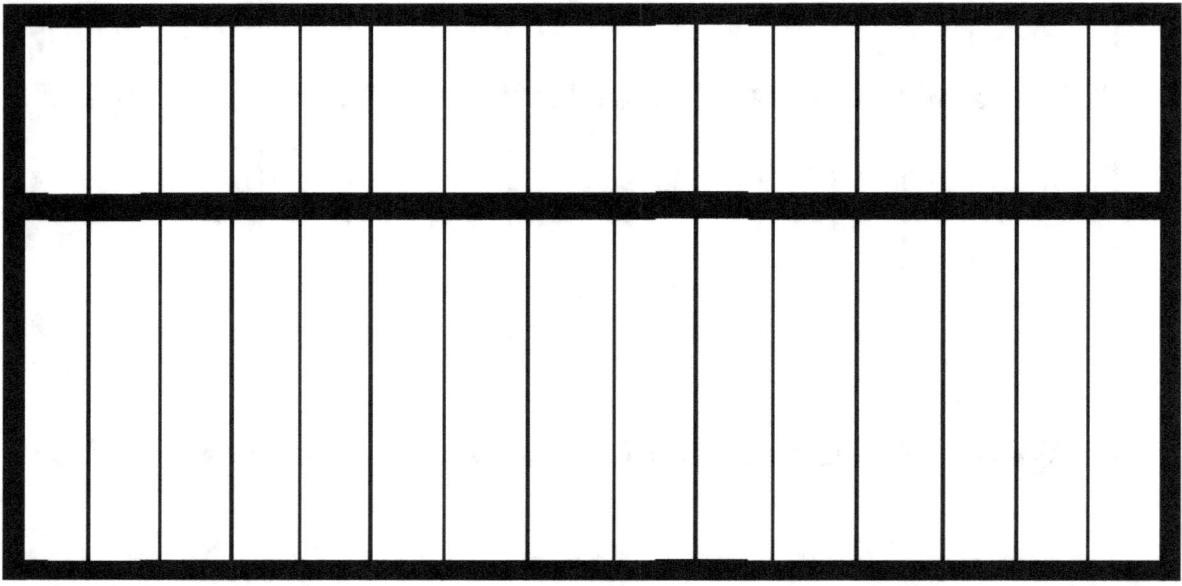

b. 304-210+190-12

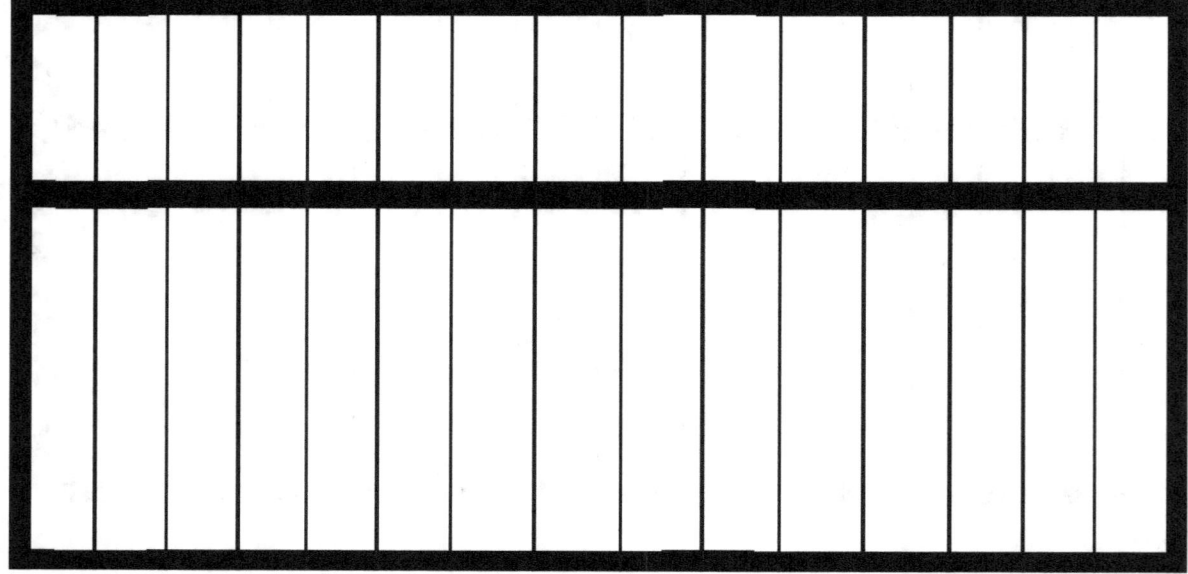

c. 213-31+180

d. 108-9+19-26

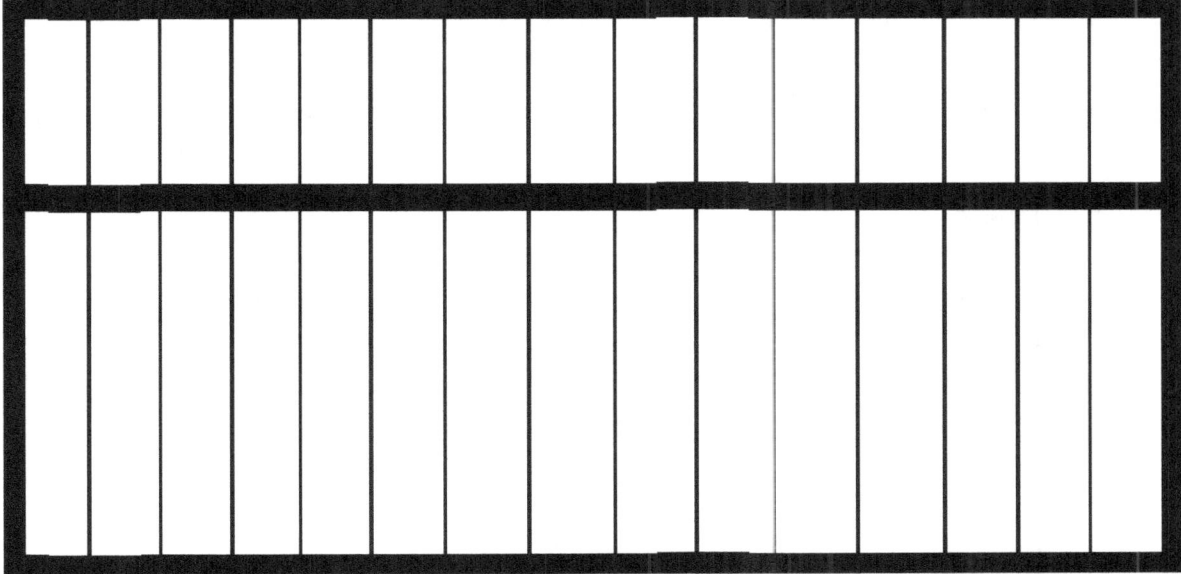

e. 999-18+54-22+75

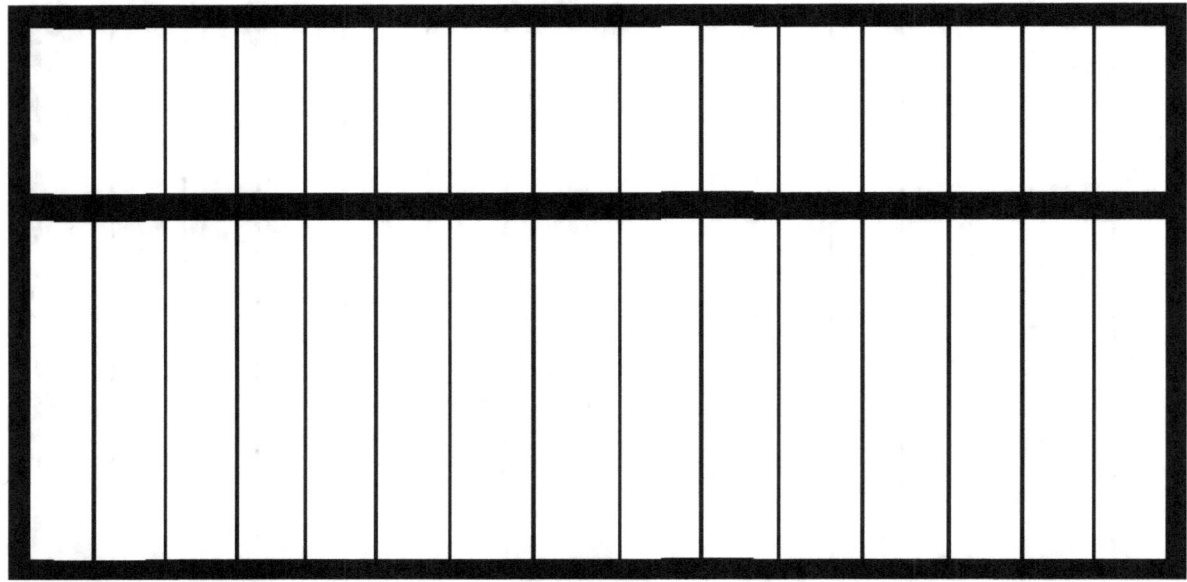

f. 211+112-121+212-2

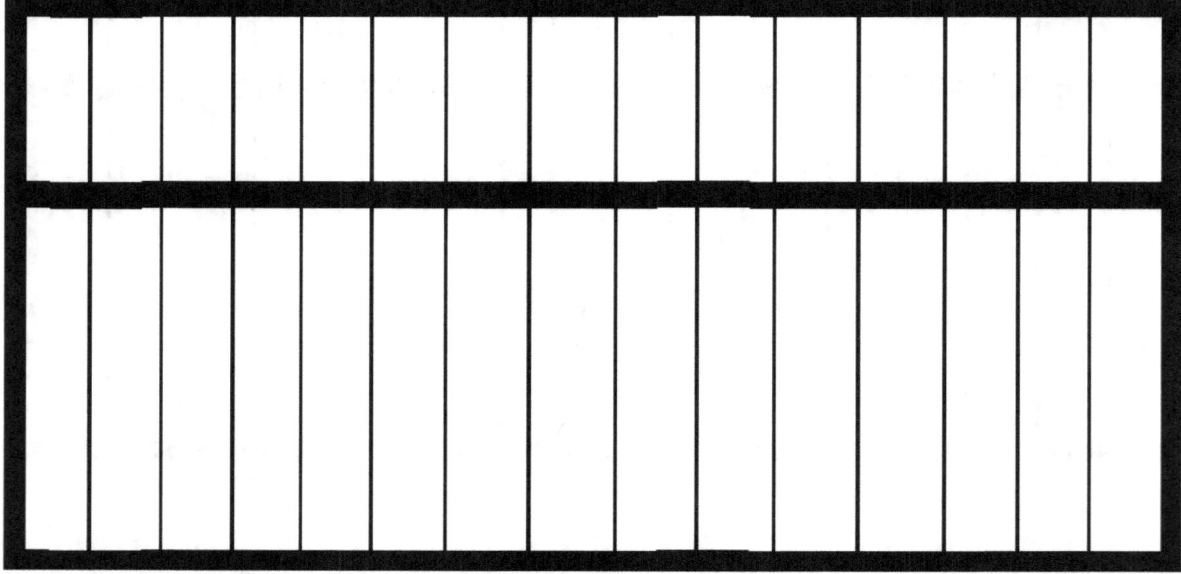

g. 309-27+91-100

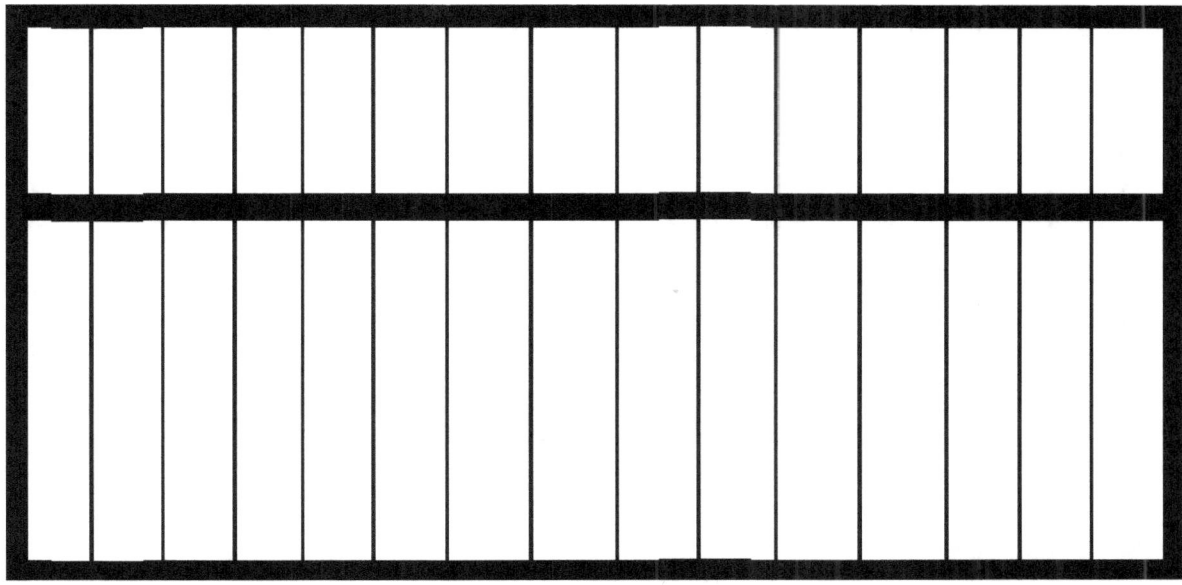

h. 209-87+91-17+20

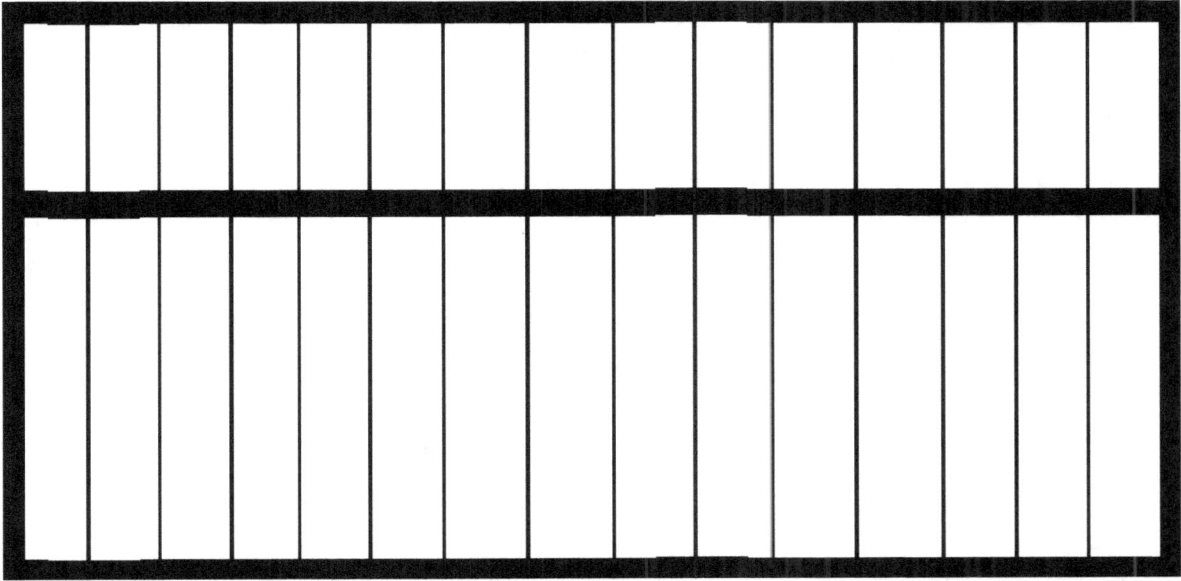

i. 120-19+109-15-22+390

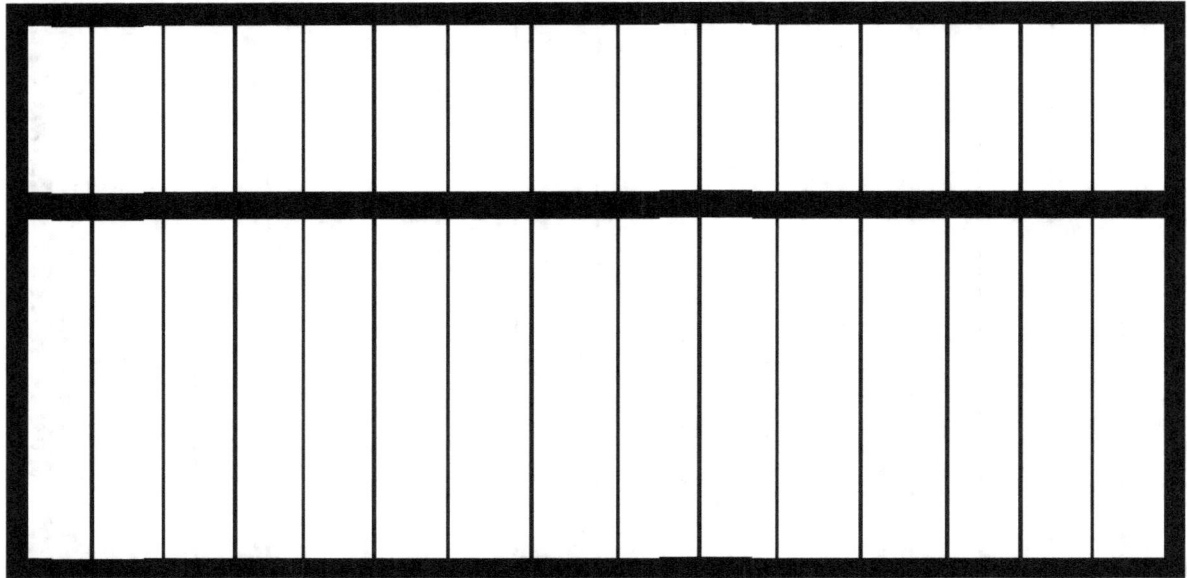

j. 3400+602-1009

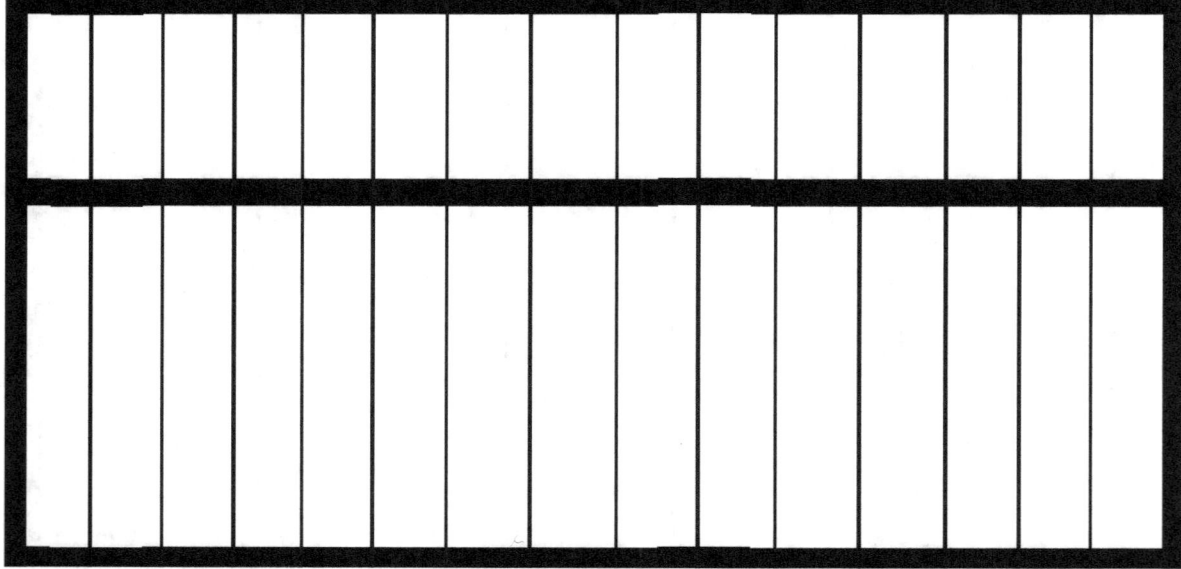

Test D, Show solution beads in final position.

Add a bar.

a. 16 x 7

b. 20 x 9

c. 13 x 4

d. (65 - 33) x 19

e. 12/4 x 64/8

f. 219 / 3

g. 33 x 47 +9 -18 / 7   *convert remainders to 2 decimal places*

h. (65 X 88) / 43

i. 23+19+37.9 + 3.5 - 12

j. $ .68 + $26.50 + $208 - $45.01

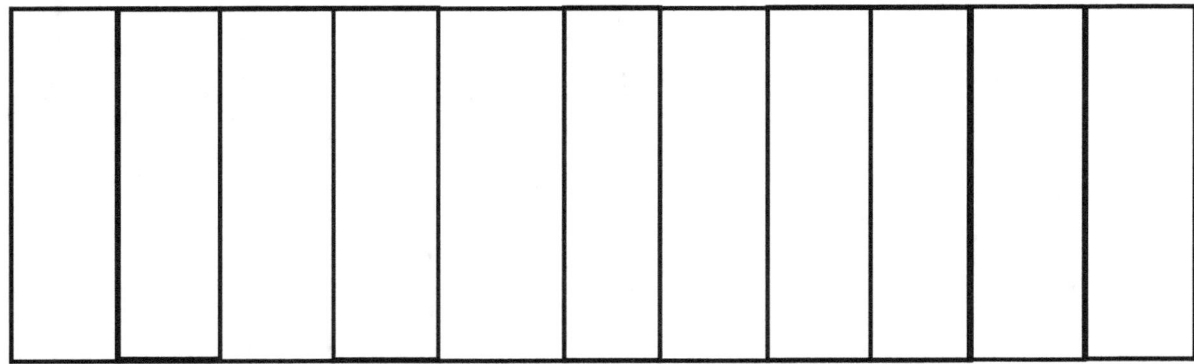

Complete all tests and then check the answers in this book, Subtract 2.5% for each wrong answer. An answer is wrong if any part of the answer is wrong. The answer may be anywhere on the Abacus, because you determine where the decimal point would be. There is no credit for completing more than what is asked for but you will give yourself an honorable mention! Add 2% for each wrong answer you are able to correct. If you score less than 75%, you need to start this book and all the work again.

100% Abacus Master

80% Abacus Expert

Your Certificate is at the back of this book.

Let me know if you enjoyed learning how to use the Chinese Abacus.

Yours in Learning and Teaching, Joe Salazar

Jws345@hotmail.com

Intentionally left blank

Test A, Answers

a. 2 2 + 3 4 =   5 6

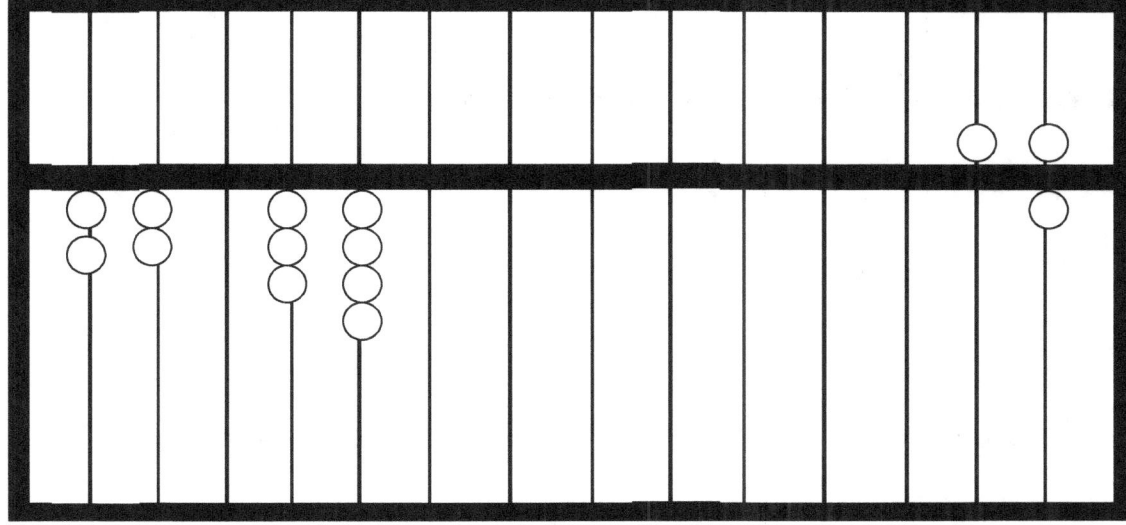

b. 1 6 + 2 6 =   4 2

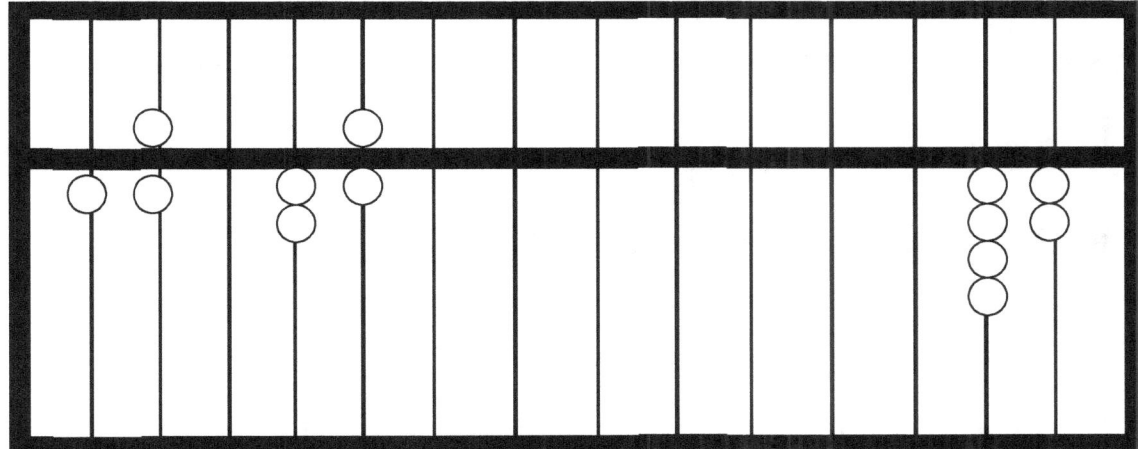

c. 104+220 =   3   2   4

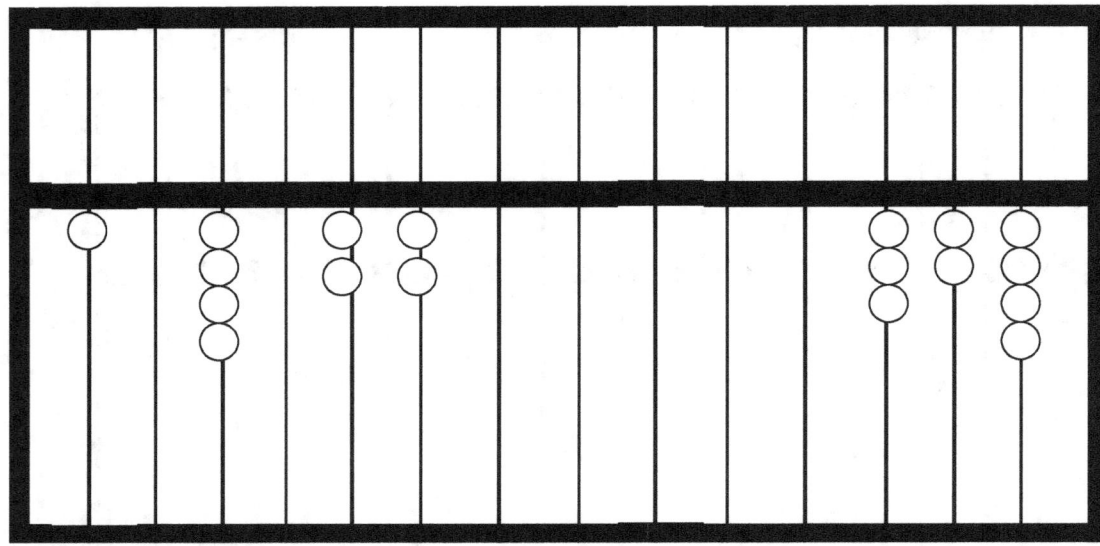

d. 30+70 = 100

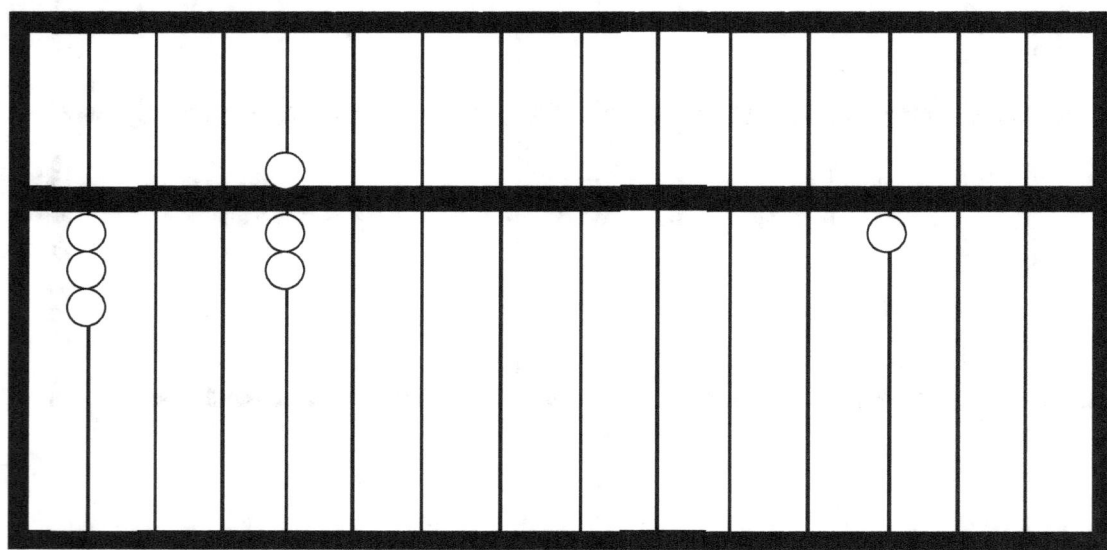

e. 4004+2030 = 6 0 3 4

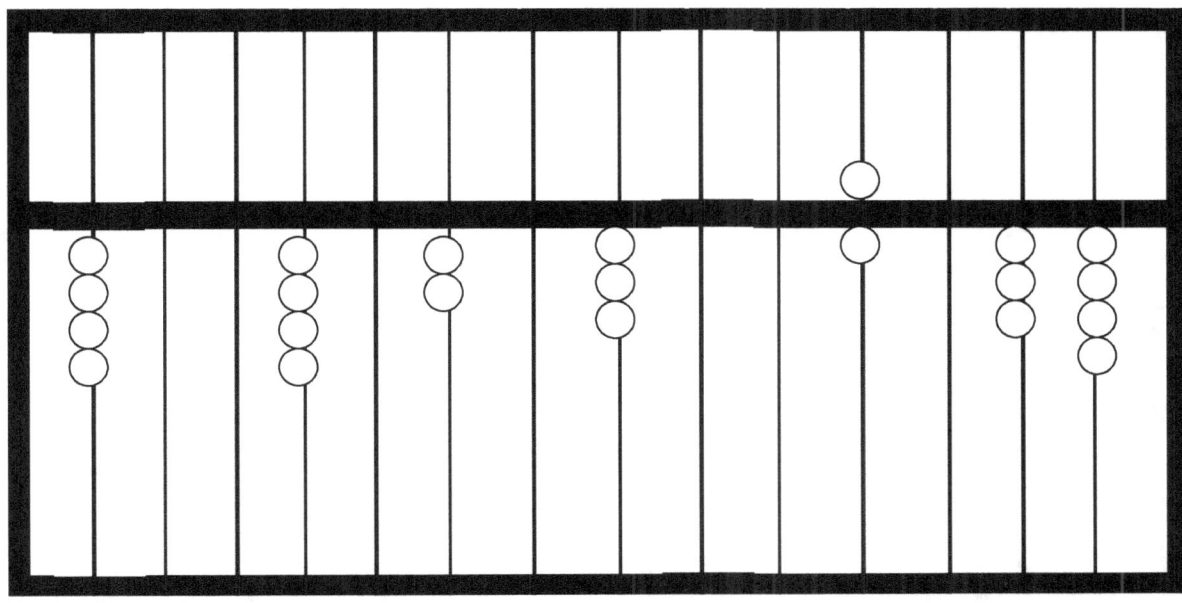

f. 180 + 810 = 990

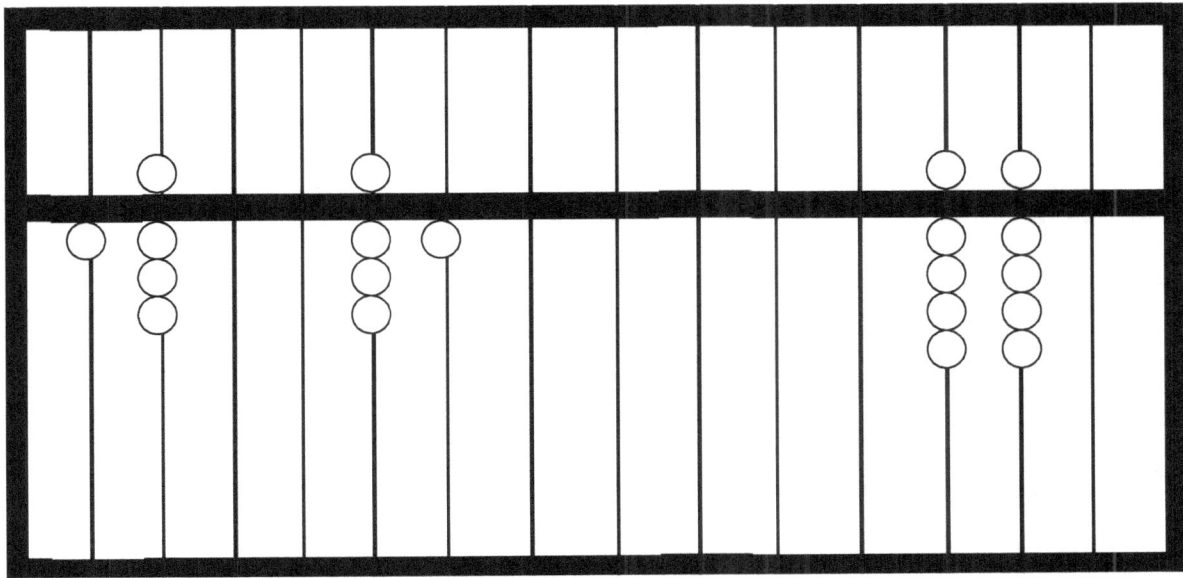

g. 999+499 = 1,498

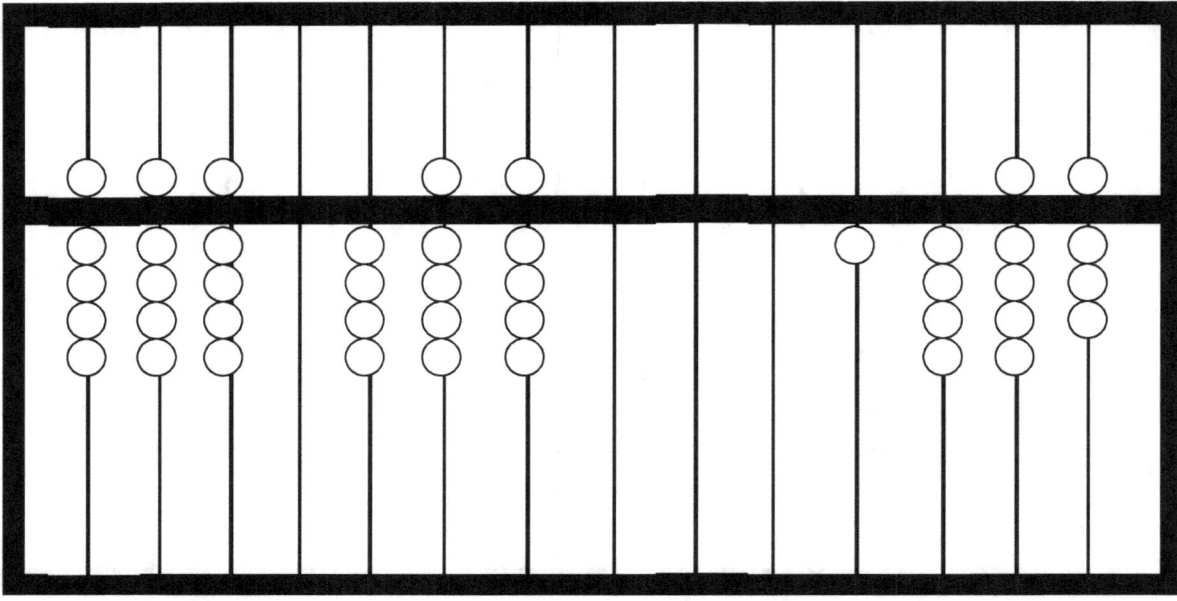

h. 115+15 = 130

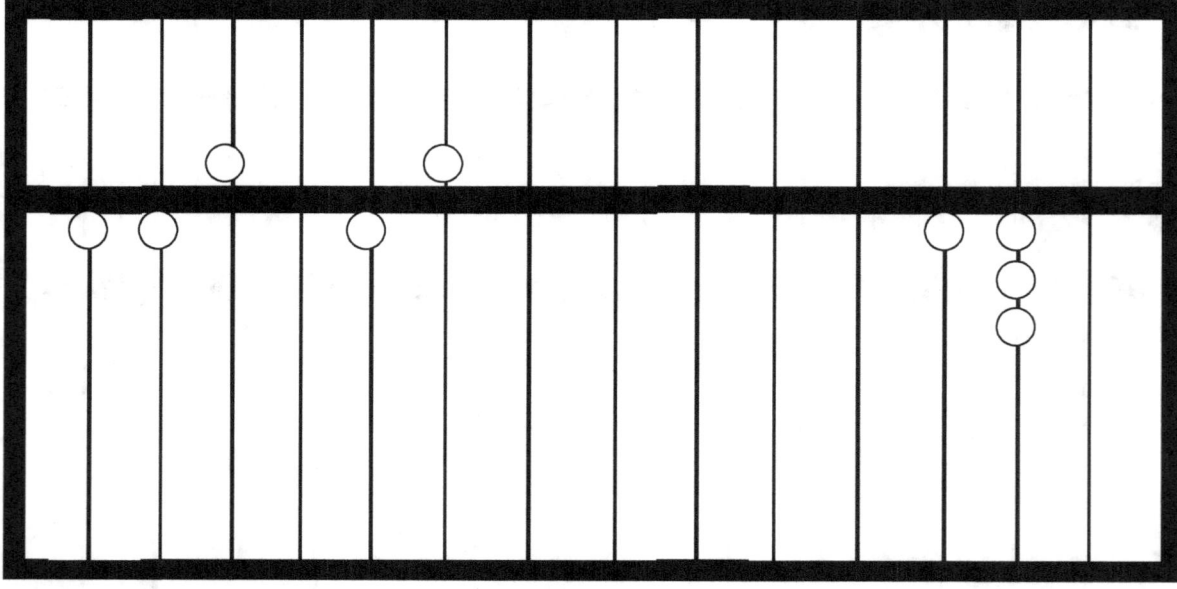

i. 203 + 219 = 422

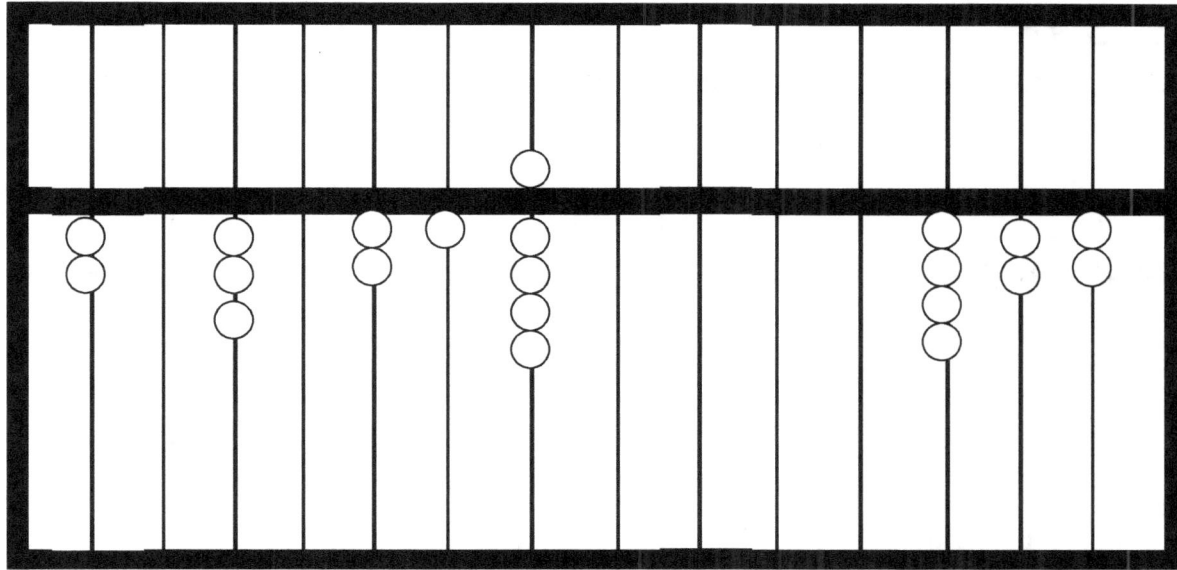

j. 4009 + 9090 = 13, 099

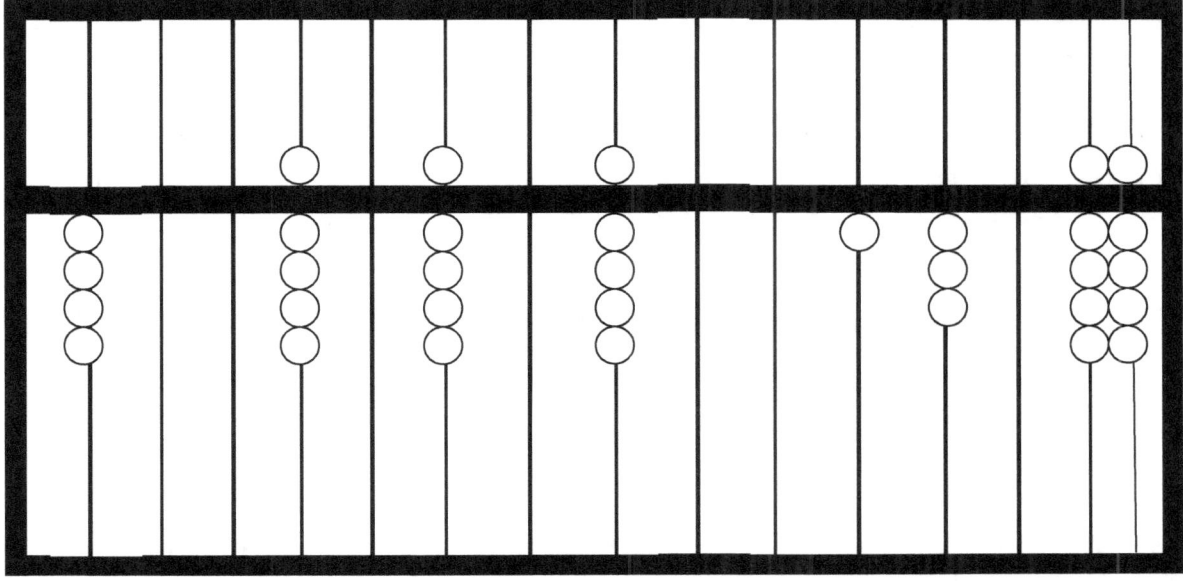

Test B, Answers

a. 35-18 = 17

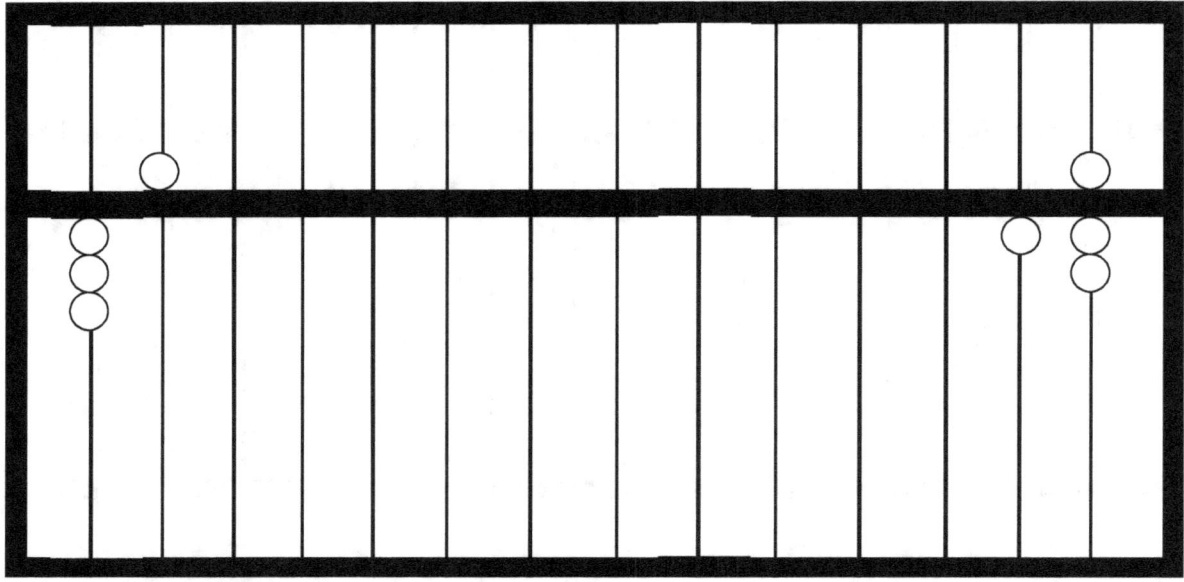

b. 355-28 = 327

c. 298-120 = 178

d. 507-203 = 304

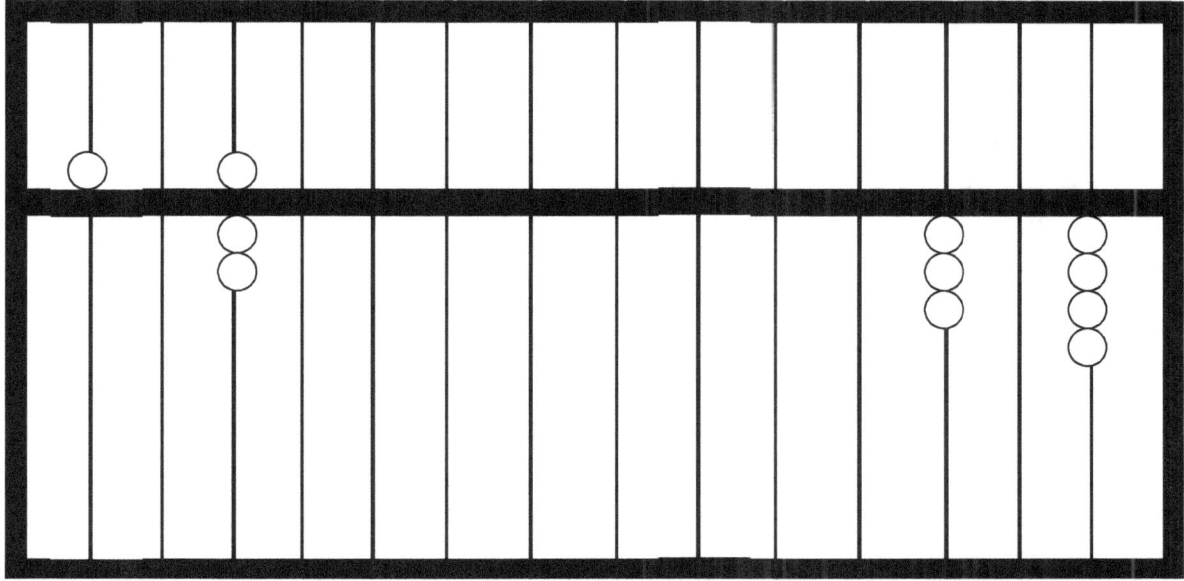

e. 3001-2003= 998

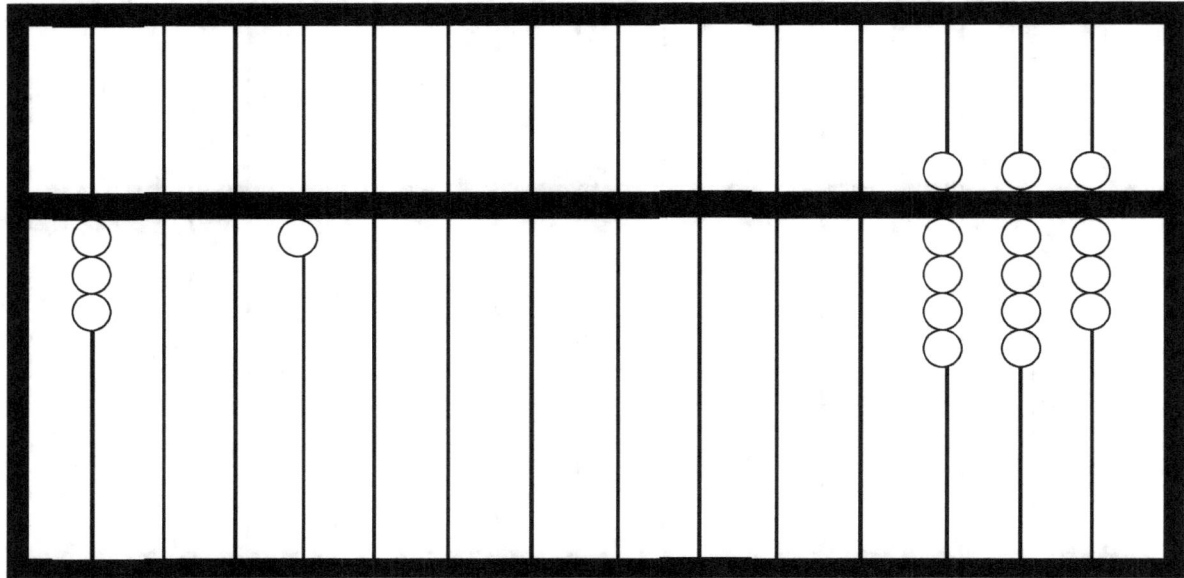

f. 100-97= 3

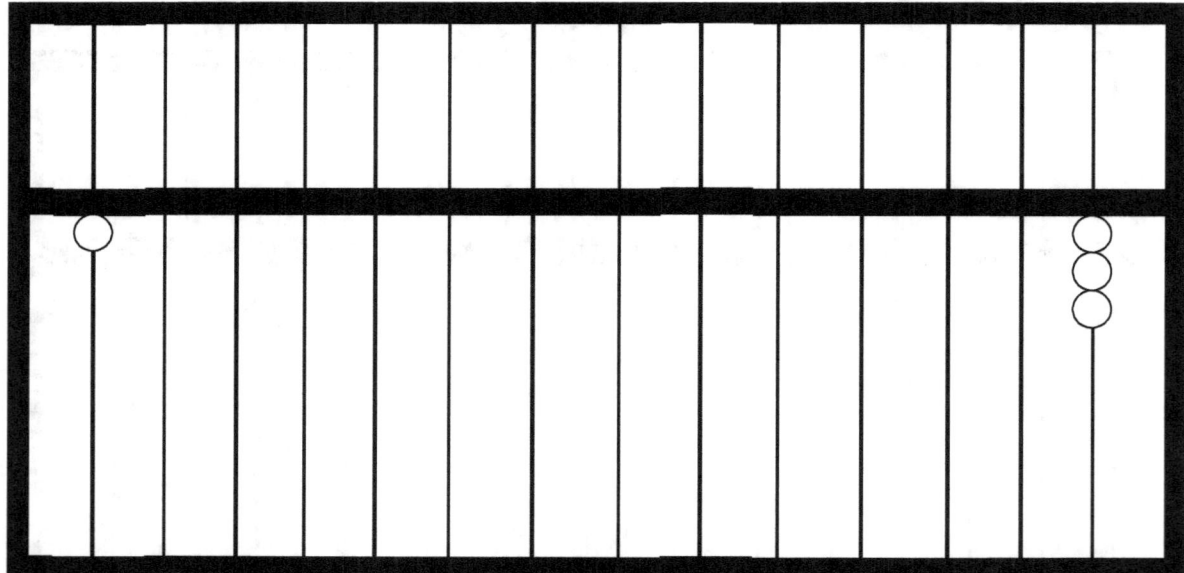

g. 222-199 = 23

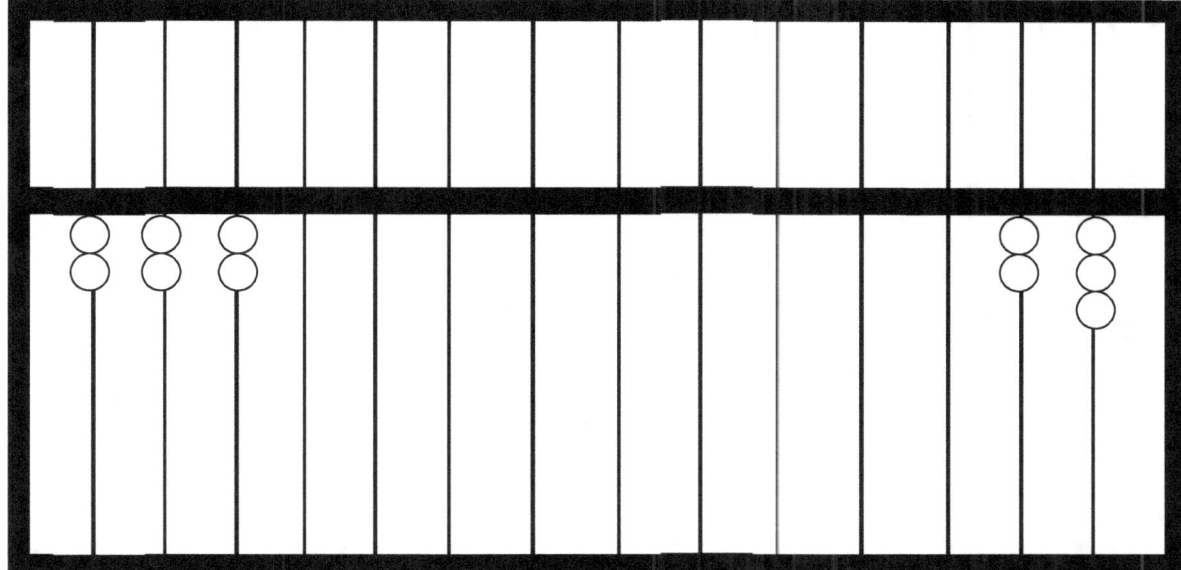

h. 606-284 = 322

i. 903-605 = 298

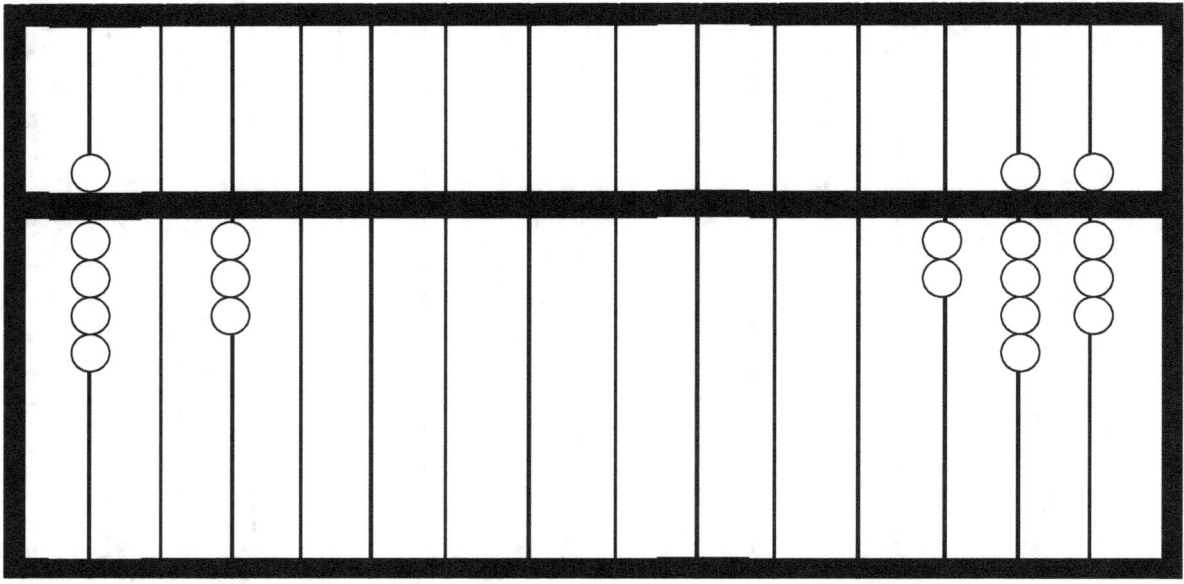

j. 505-209 = 296

Test C, Diagram only the solution calculated on abacus

a. 23+18-9+17 =49

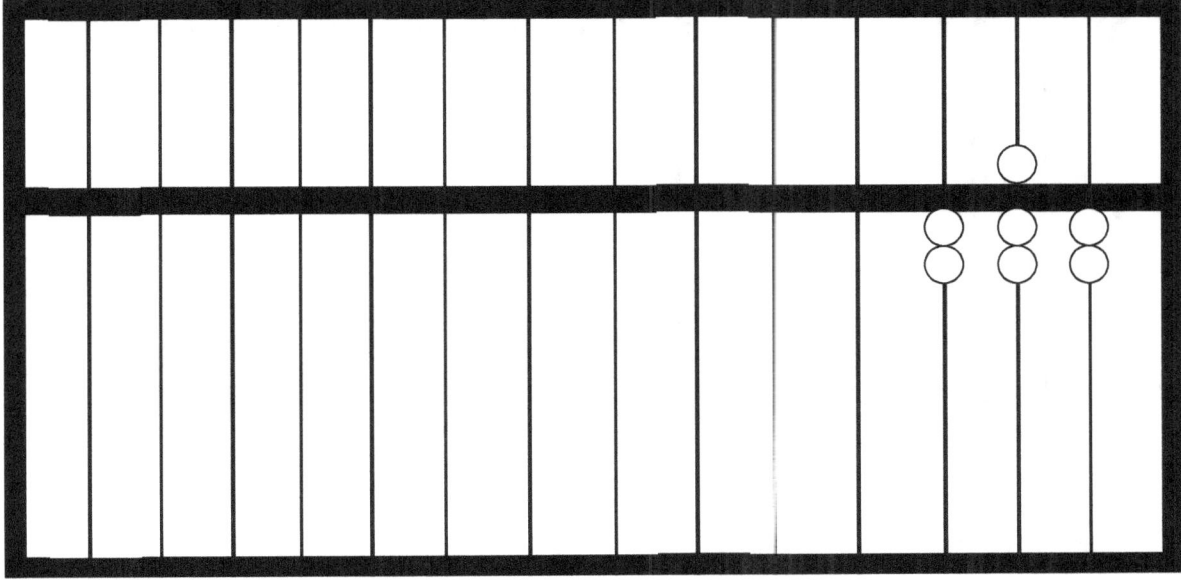

b. 304-210+190-12 = 272

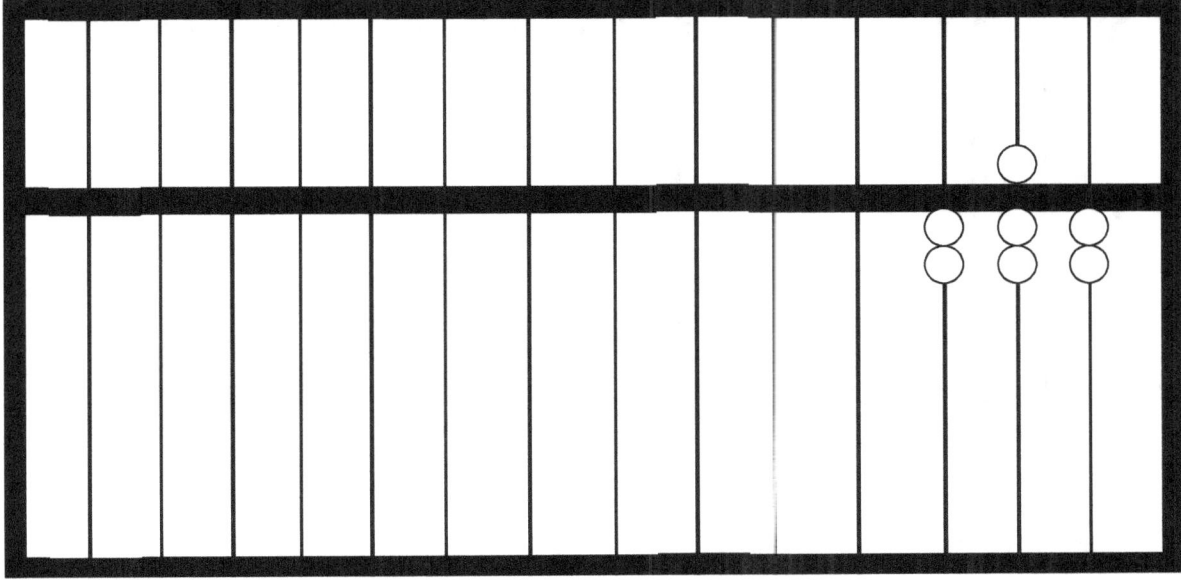

c. 213-31+180 = 362

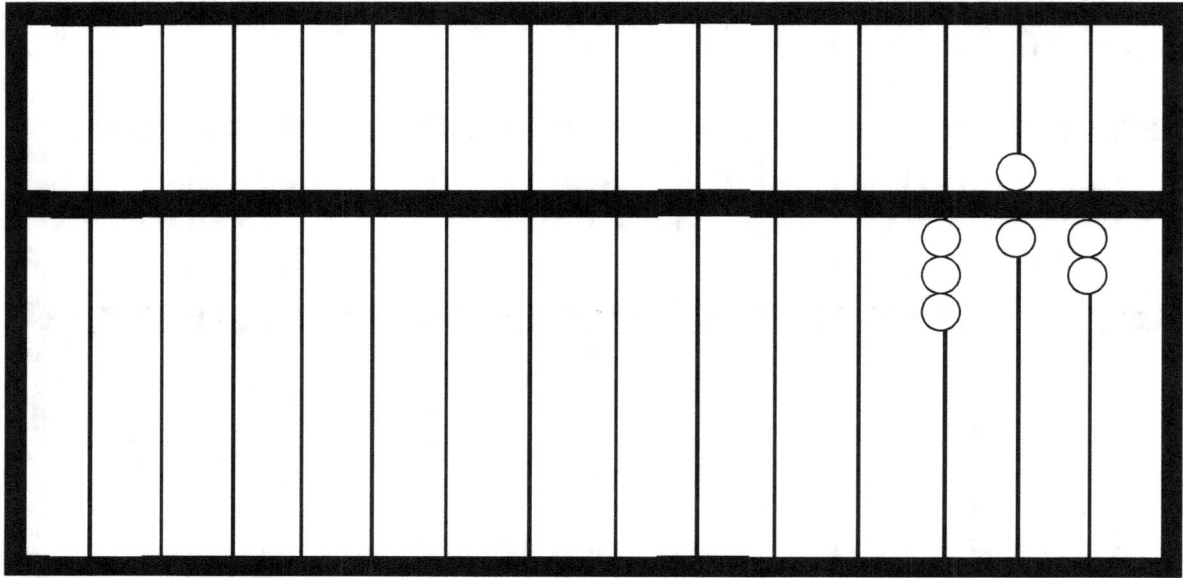

d. 108-9+19-26 = 92

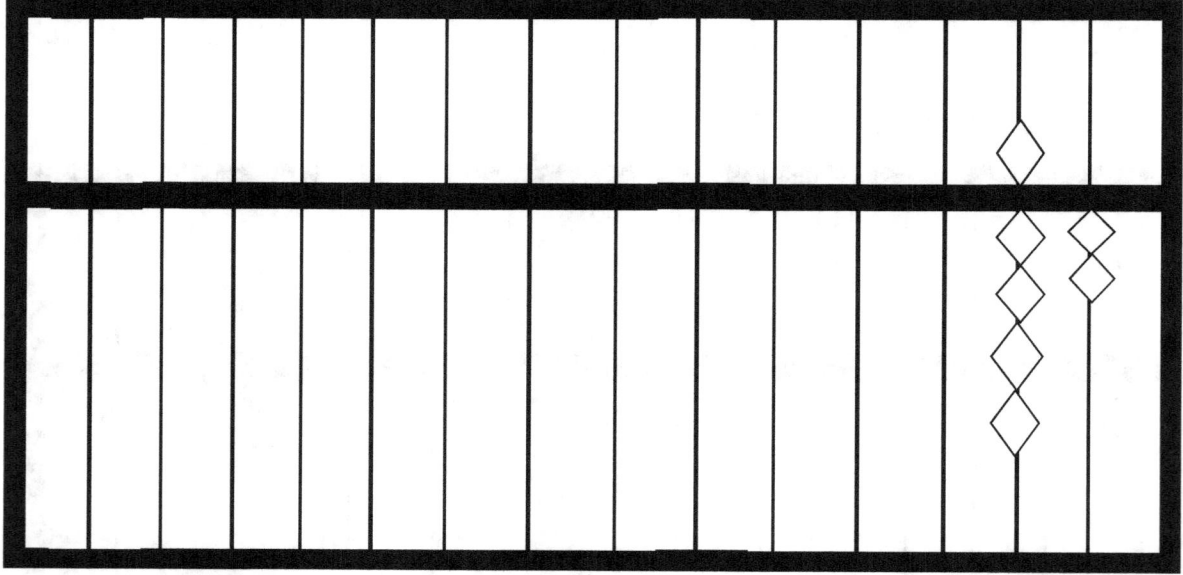

e. 999-18+54-22+75 = 1088

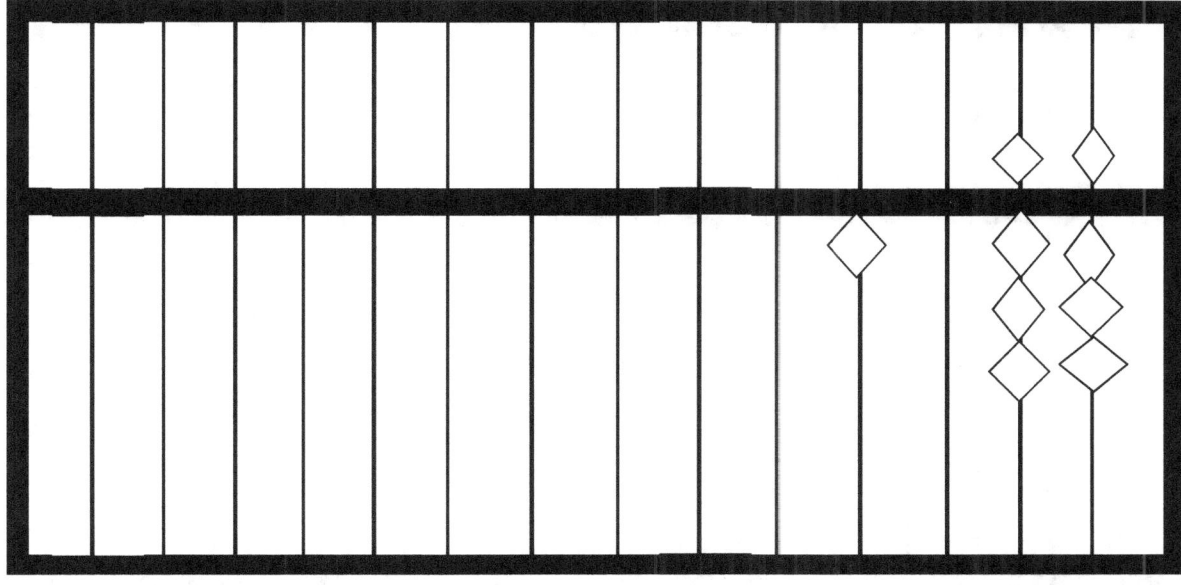

f. 211+112-121+212-2 = 412

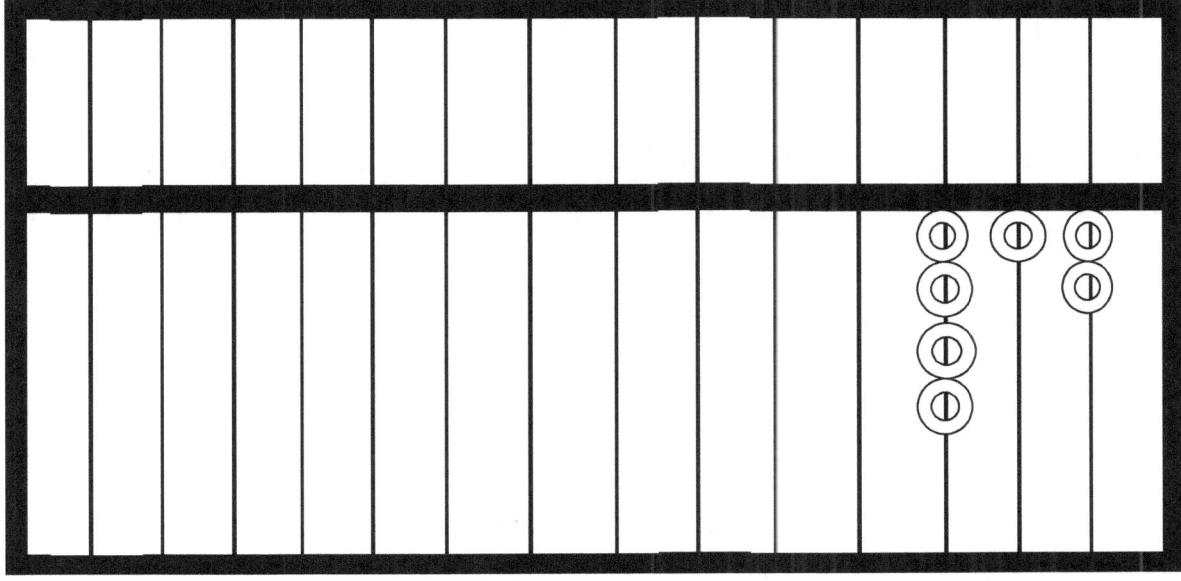

g. 309-27+91-100 = 273

h. 209-87+91-17+20 = 216

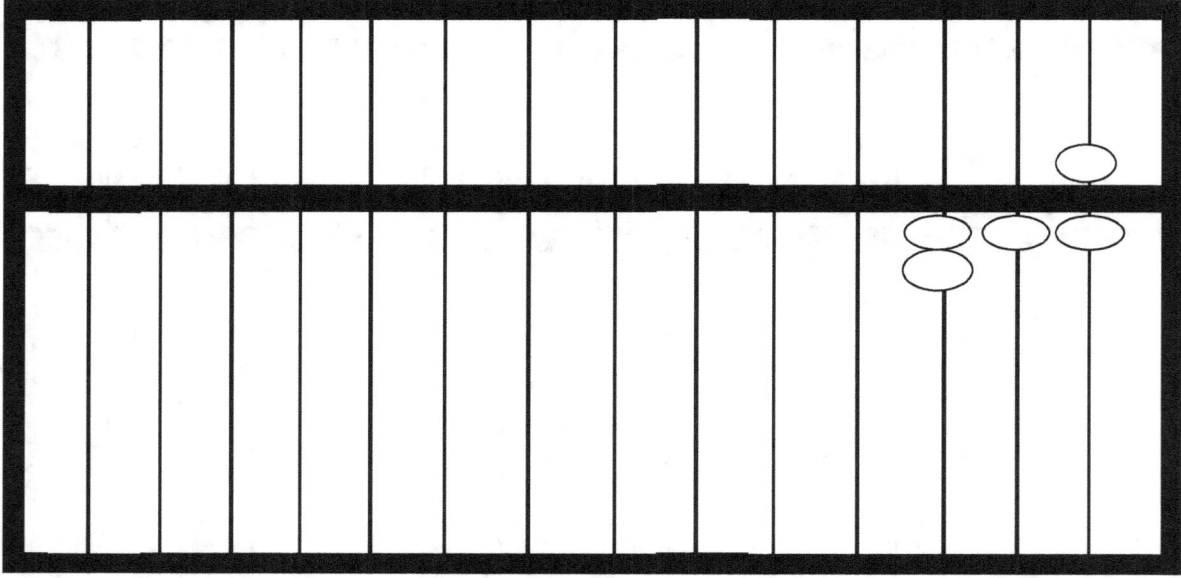

i. 120-19+109-15-22+390 = 563

j. 3400+602-1009 = 2993

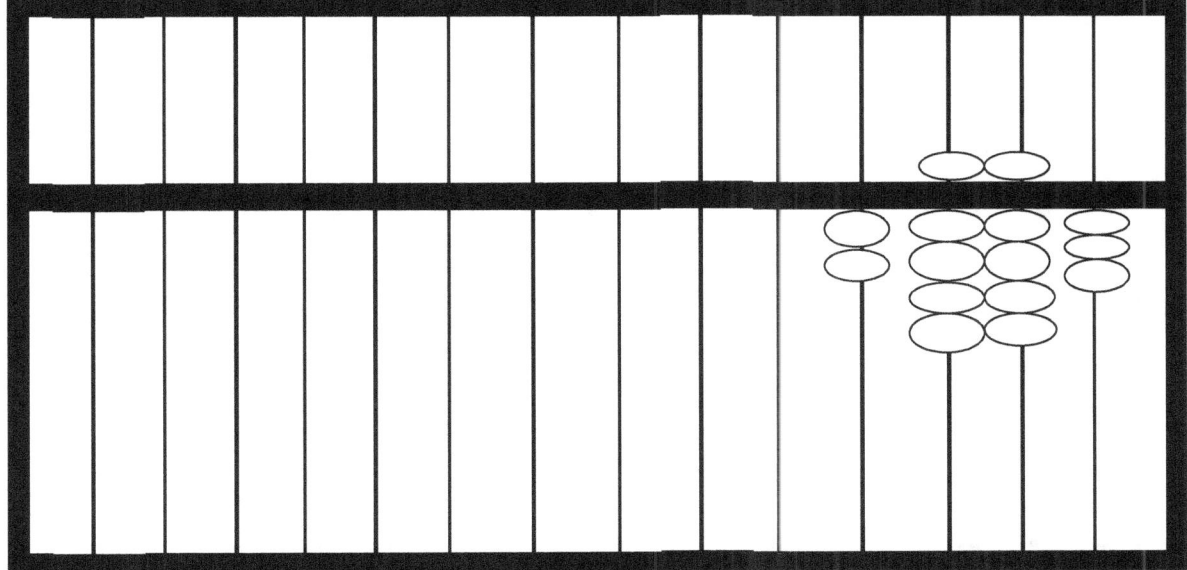

Test D, Answers

a. 16 x 7

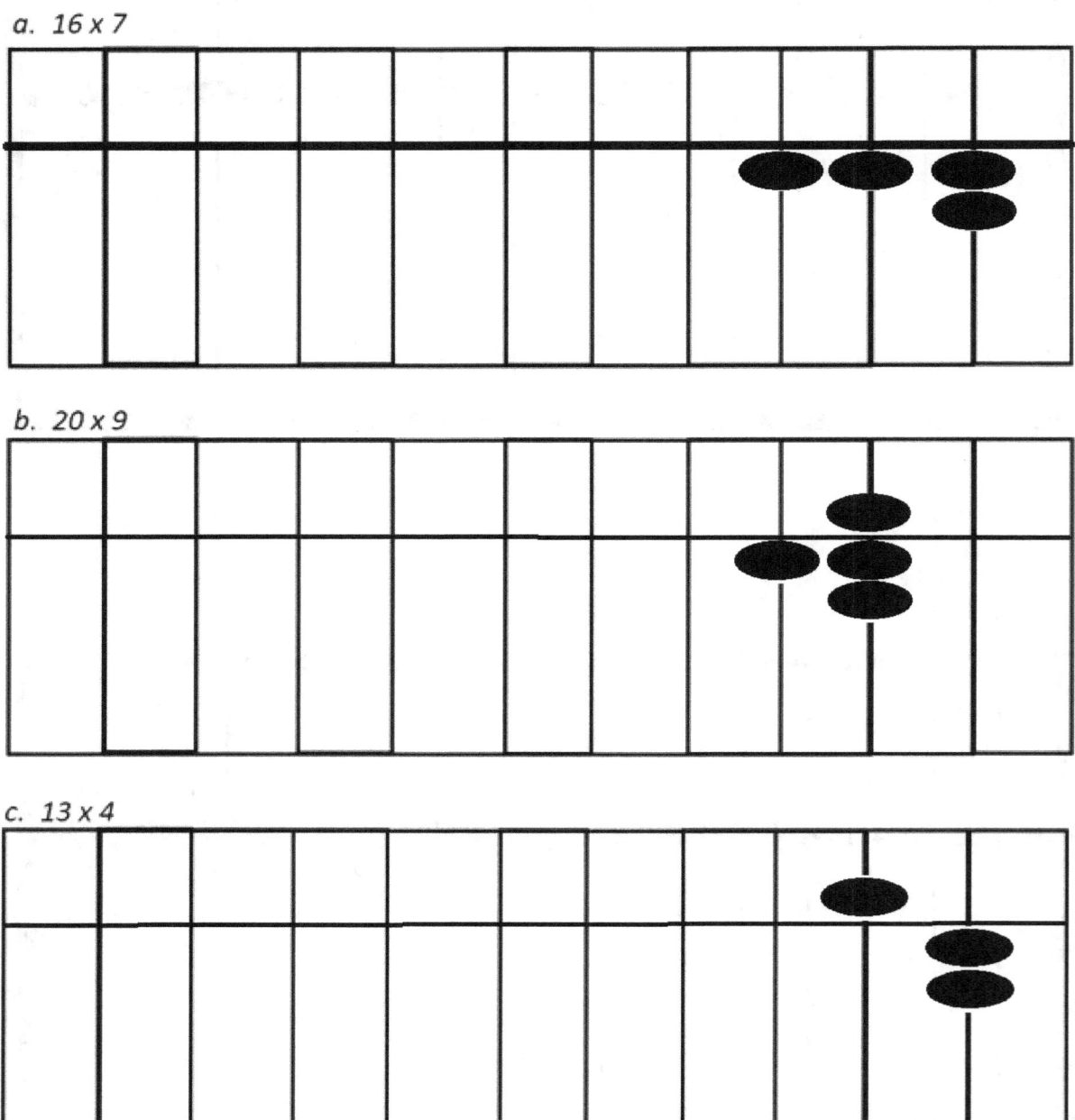

b. 20 x 9

c. 13 x 4

d. (65 - 33) x 19

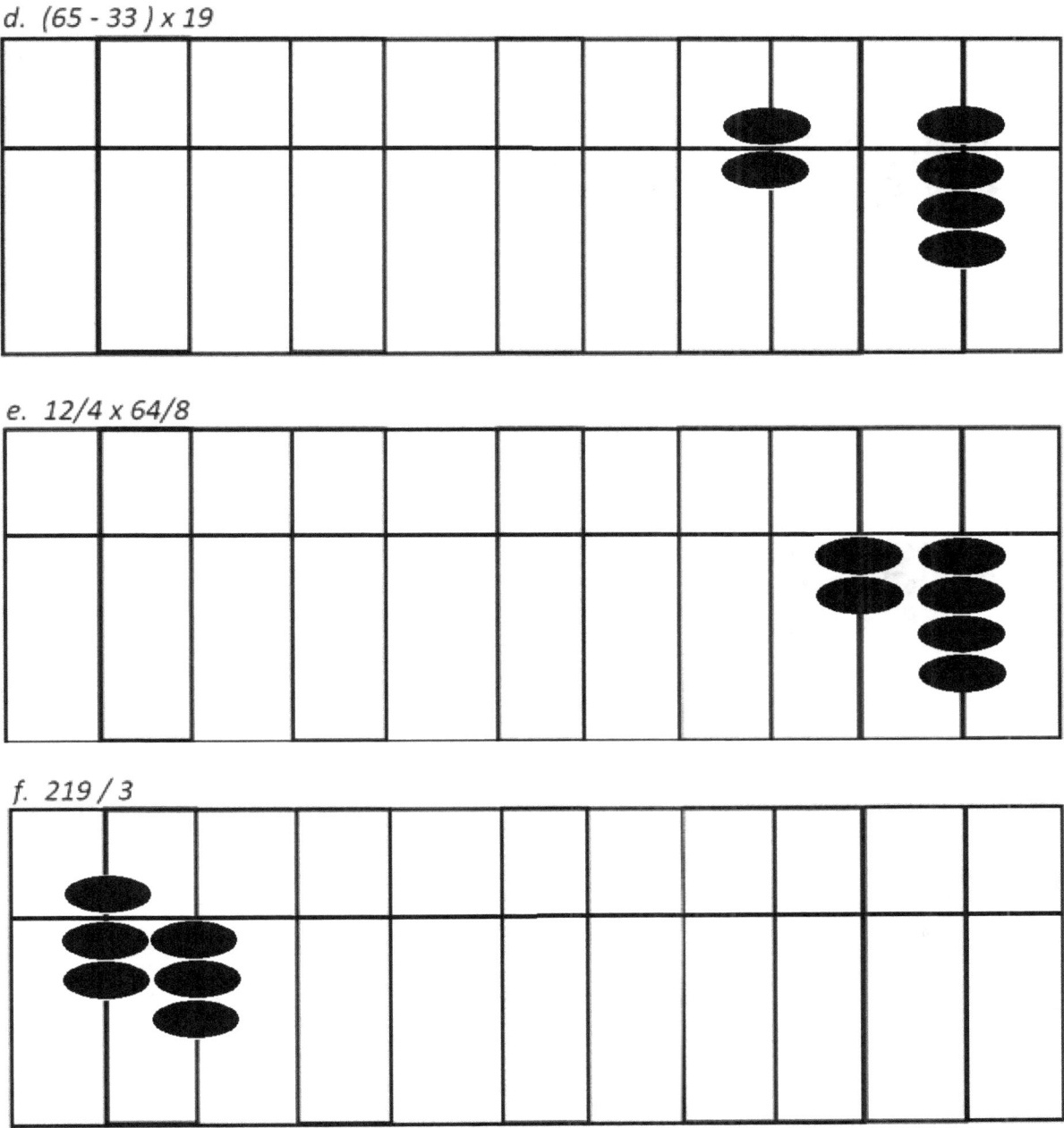

e. 12/4 x 64/8

f. 219 / 3

It is appropriate to plan a quotient from left to right because we don't know how many spaces we will need.

g. 33 x 47 +9 -18 / 7   *convert remainders to 2 decimal places*

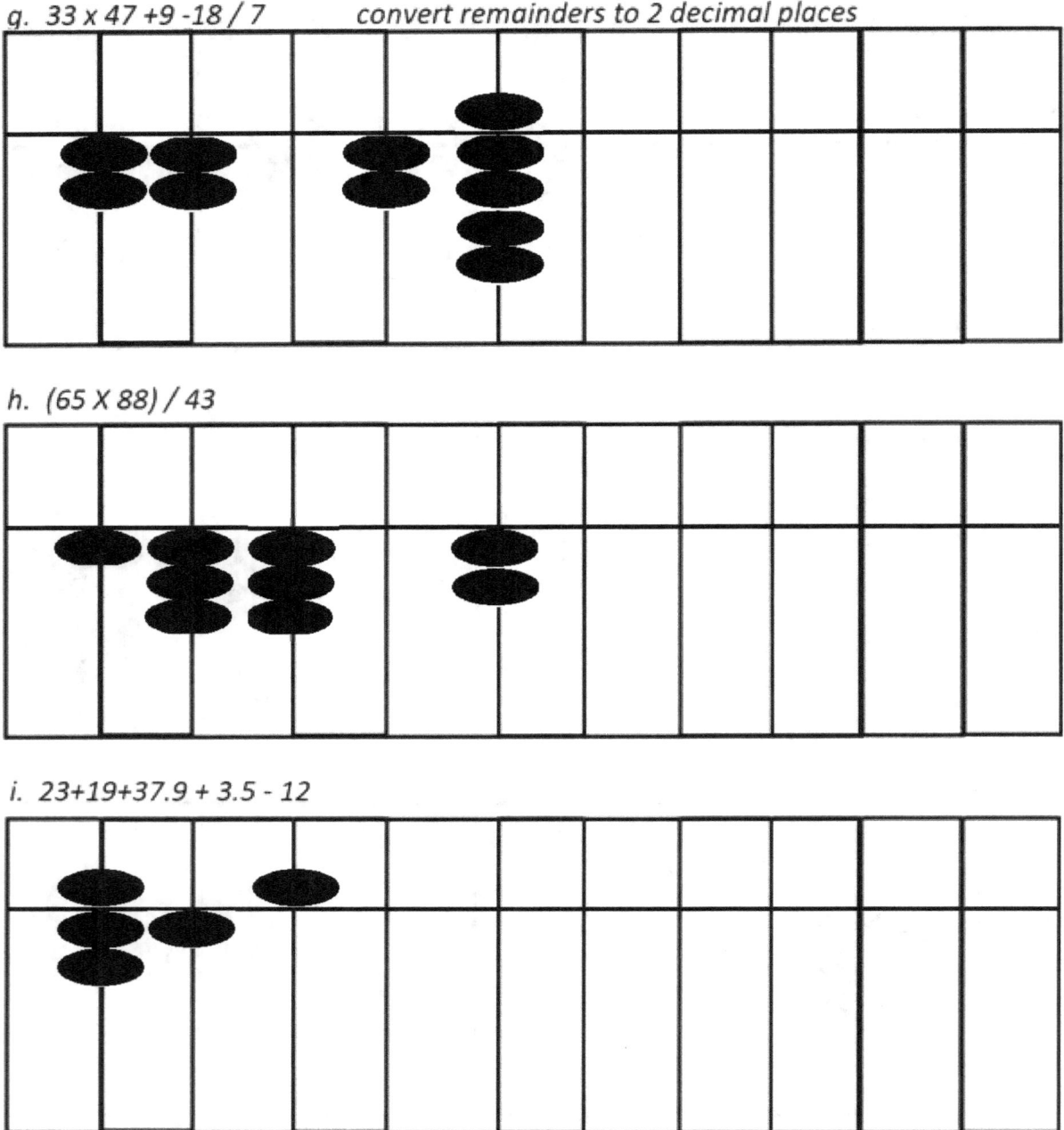

h. (65 X 88) / 43

i. 23+19+37.9 + 3.5 - 12

Congratulations you have completed 10 hours of
Abacus.

The Author

Joseph W. Salazar

Born Dec. 9, 1940 Chicago Illinois

Enlisted in the U.S. Army at age seventeen, April 1958

Assigned to the Medical Corps and served six years, four of which were in Germany. I worked in many types of Army Medical Clinics and hospitals as a Medical Attendant and Technician. After leaving the Army I lived in Salt Lake City for two years before moving to California and worked as a Vocational Nurse, Driving Instructor, Martial Arts Teacher, and Traffic Violator School Instructor, while obtaining my B. S. in Health Science at San Jose State University. I got my J. D. from Saratoga University Law School, Distance Learning Program.

My hobbies are Chess, Martial Arts, Archery and Flying Planes.

I taught the Accounting Clerk Program for Goodwill Industries for three years and operated my own Tax & Accounting business for forty years. I continue to operate my own business as a Public Accountant.

Joe Salazar

Prunedale California, June 2023

Intentionally left blank

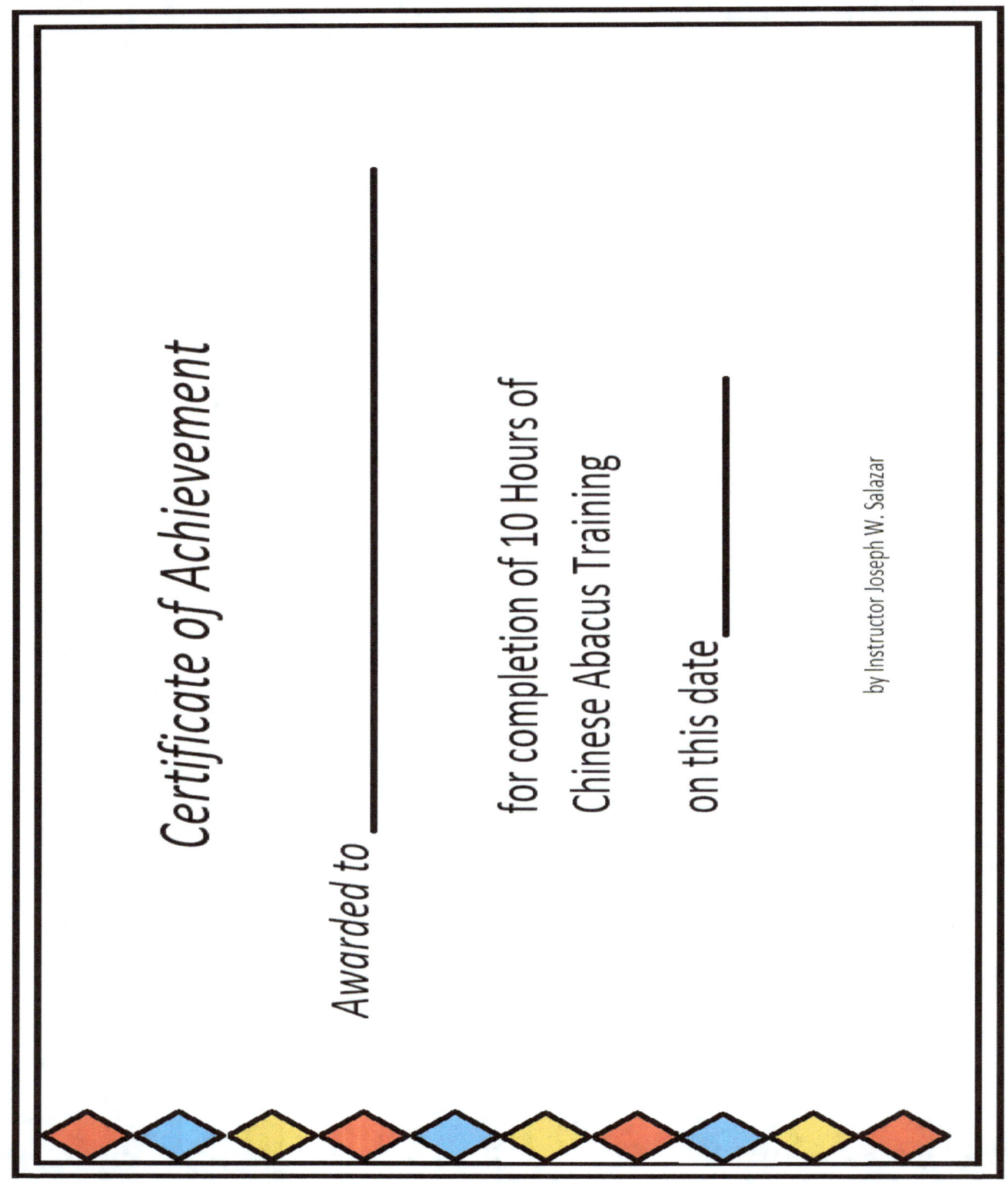

Certificate of Achievement

Awarded to _____

for completion of 10 Hours of
Chinese Abacus Training

on this date _____

by Instructor Joseph W. Salazar